GROW YOUR SOIL!

HARNESS THE POWER OF THE SOIL FOOD WEB TO CREATE YOUR BEST GARDEN *EVER*

By Diane Miessler

The mission of Storey Publishing is to serve our customers by publishing practical information that encourages personal independence in harmony with the environment.

Edited by Carleen Madigan and Sarah Guare
Art direction and book design by Michaela Jebb
Text production by Jennifer Jepson Smith
Illustrations by © Kristyna Baczynski

Text © 2020 by Diane Miessler

All rights reserved. Hachette Book Group supports the right to free expression and the value of copyright. The purpose of copyright is to encourage writers and artists to produce the creative works that enrich our culture. The scanning, uploading, and distribution of this book without permission is a theft of the author's intellectual property. If you would like permission to use material from the book (other than for review purposes), please contact permissions@hbgusa.com. Thank you for your support of the author's rights.

The information in this book is true and complete to the best of our knowledge. All recommendations are made without guarantee on the part of the author or Storey Publishing. The author and publisher disclaim any liability in connection with the use of this information.

The publisher is not responsible for websites (or their content) that are not owned by the publisher.

Storey books may be purchased in bulk for business, educational, or promotional use. Special editions or book excerpts can also be created to specification. For details, please contact your local bookseller or the Hachette Book Group Special Markets Department at special.markets@hbgusa.com.

Storey Publishing
210 MASS MoCA Way
North Adams, MA 01247
storey.com

Storey Publishing is an imprint of Workman Publishing, a division of Hachette Book Group, Inc., 1290 Avenue of the Americas, New York, NY 10104. The Storey Publishing name and logo are registered trademarks of Hachette Book Group, Inc.

ISBNs: 978-1-63586-207-2 (paperback); 978-1-63586-208-9 (ebook)

Printed in China by R. R. Donnelley on paper from responsible sources
10 9 8 7 6 5 4 3

LIBRARY OF CONGRESS CATALOGING-IN-PUBLICATION DATA

Names: Miessler, Diane, author.
Title: Grow your soil! : harness the power of the soil food web to create your best garden ever / by Diane Miessler.
Description: [North Adams, Mass.] : Storey Publishing, 2020. | Includes bibliographical references and index. | Summary: "Diane Miessler presents the science of soil health and shares the techniques she has used — including cover crops, constant mulching, and compost tea — to create and maintain rich, dark, crumbly soil that's teeming with life" — Provided by publisher.
Identifiers: LCCN 2019051715 (print) | LCCN 2019051716 (ebook) | ISBN 9781635862072 (paperback) | ISBN 9781635862089 (ebook)
Subjects: LCSH: Soil management. | Soils. | Gardening.
Classification: LCC S591 .M65 2020 (print) | LCC S591 (ebook) | DDC 631.4—dc23
LC record available at https://lccn.loc.gov/2019051715
LC ebook record available at https://lccn.loc.gov/2019051716

DEDICATION

To my amazing kids, Abe and Zoe, who grew
up with my quirks and love me anyway.

To Jess and Nate, my bonus kids.

To Walter, who learned with me
how to be happily married.

To my dad, who taught me most of what
I know about gardening and about love.

To my grandkids Maggie, Linc, and Charlie,
who make the world worth saving.

ACKNOWLEDGMENTS

I'd like to thank my family and friends who have been cheering me on for years, especially Linda Miessler, Glennda Chui, Beth Leydon, Katie McCamant, Catherine Allen, Mary Ann Johnston, Jeff Wright, and Molly Fisk.

And thank you to Tom Durkin for editing my rough draft and making it less incomprehensible. I couldn't have done it without you!

CONTENTS

Foreword by Elaine R. Ingham — viii

Beauty, Magic & Tomatoes:
Why I Garden, and How the Soil Food Web Helps — 1

10 ~~Commandm~~ Suggestions for Creating Healthy, Living Soil — 4

Making Soil: Building a "House" — 6

1 THE SKINNY ON SOIL
What Makes Soil Good & How Does It Get That Way? — 8
What *Is* Soil? — 10
What Makes Good Soil Good? — 11
What Good Is Good Soil? — 14

2 BUILDING A HOUSE
Start with the Roof (Mulch & Cover Crops) — 20
Mulch — 22
Cover Crops — 27

3 BUILD THE WALLS
Organic Matter & the Soil Food Web That Sticks It Together — 40
Soil Organic Matter — 42
The Soil Food Web: A Gardener's Best Friend — 44
Biodiversity 101: Microbes — 46
Biodiversity 102: Algae, Bugs, Worms, and More — 62

4 INSTALL VENTILATION, PLUMBING & A NICE PANTRY
No-Till Growing, Paths & CEC — 70
Step Away from the Rototiller! — 72
Make Paths — 75
What is CEC? — 77
How to Increase CEC — 78

5 FEED THE INHABITANTS
Photosynthesis, Minerals & Soil Testing — 84
Photosynthesis: Empty but Necessary Calories — 86
Macronutrients & Micronutrients — 87
To Test or Not to Test — 95

6 COMPOST & COMPOST TEA
It's Not Rocket Science — 106
Benefits and Tools — 108
Is It Compost or Is It Mulch? — 110
How to Build a Compost Pile — 111
Maintaining a Compost Pile: It Needs Your Love — 114
When Is It Done? — 117
Growing Pains — 118
Fungally or Bacterially Dominated? — 119
Worm Bins — 121
Frequently Asked Questions, Answered — 123
Compost Tea for a Non-Rocket Scientist with a Bucket — 128

7 WHERE DO I START?
Building a Garden That Feeds Itself — 132
Step 1: Pick a Spot — 134
Step 2: Get Water to It — 134
Step 3: Decide Where You'll Walk — 136
Step 4: Soften the Soil — 136
Step 5: Pull Stuff Up, Throw Stuff Down — 137
Step 6: Mulch — 137
Step 7: Plant Stuff — 138

8 HOUSEWORK & HOME REPAIRS
How *Not* to Make Your Neighbors Hate You — 144
Maintenance = Mulch — 146
Putter, Don't Work — 146

Glossary — 156
Bibliography — 159
Metric Conversions — 161
Index — 162

FOREWORD
By Elaine R. Ingham

Reading this book is such a joy. With Diane's guidance, gardening becomes an ongoing observation of the natural interactions between the many organisms that abound in a healthy garden. There's so much to look for, experience, and comprehend. To Diane, gardening is an ongoing conversation held with your plants.

Diane's descriptions of the different groups of organisms in the garden made me smile and, on occasion, laugh out loud. This is the way science should be presented: as a perpetual discovery of the organisms that inhabit your soil and how best to handle them to help your plants grow. No more "if it moves, kill it!" That approach is too heavy-handed, too extreme. This book teaches you subtle management techniques for growing good food and beautiful landscapes.

Nature manages soil life very cleverly, with a minimum of fuss, muss, and mistakes. We need to understand how nature does this, and then work *with nature*. It's amazing how easy it is to grow any plant when you understand what it needs. Cooperation and networking are the dominant forces leading to greater productivity — not destruction. Nuking your "enemies" — diseases, pests, and parasites — is not the way to exist. Down that road is a horrible future.

People who climbed on the "toxic chemical wagon" were told it was the only way to grow enough food to feed a starving world. That was a flat-out lie. People were also told that inorganic fertilizers and pesticides don't harm soil life when in fact, these toxins destroy nearly all the beneficial organisms in the soil and on plant surfaces. If you've damaged the life in your soil and on your plants, this book can help you replenish what is missing. Using the techniques Diane describes, you can grow equal or greater yields of highly nutritious foods with less cost.

You will learn how to work with nature to support beneficial organisms, so that if a not-so-wonderful creature — like a disease, pest, or parasite — comes looking for a place to hang out, it won't be able to find a comfortable spot and so will move on. These problem organisms are messengers from nature that something is wrong in your soil. Treat the cause, not the symptom — don't kill the messenger!

Diane's humor and tongue-in-cheek joy make this book so fun to read. Over and over again, she revels in the downright silliness that occurs in the wilds of your garden, in your soil, and between some of the craziest critters you never knew existed. Let this book help you go on safari in your own wild kingdom, just beyond your back door. And if you don't yet have a garden worthy of going on safari in, then this book will help you build one teeming with the beneficial life that should be there.

If you understand how nature does things, you do not need toxic chemicals or inorganic fertilizers, nor do you need to be constantly disrupting the soil. Once you take the first step on the path to soil health, take a second step, a third, a fourth . . . This book can help lead you down that garden path, to a delightful place that promotes health *and* keeps your soil critters happy!

BEAUTY, MAGIC & TOMATOES

Why I Garden, and How the Soil Food Web Helps

My first gardening memory is of planting wrinkly gray seeds in orange juice cans with my dad (best dad EVER, by the way, and the likely source of my gardening bug).

I watched with a sort of proprietary amazement as those seeds sprouted into sturdy green shoots, and then grew into the plant that still emanates my favorite smell: sweet peas, in an exuberant mix of pinks, blues, and purples. I didn't know how wrinkly gray seeds could do that. I still don't; what I do know is that I still love making it happen.

I garden because gardening makes ugly places beautiful, grows food, and, perhaps most important, because plants amaze me.

MAKING MAGIC. Gardening is the closest thing to magic I know. Well, that and babies — where do they come from? I know, I know: "Sometimes, when a man and a woman love each other very much . . ." I've read the literature. The literature, however, doesn't explain how an invisible glob of protoplasm turns into a thinking, reasoning human being with its own spirit and personality. Or how a seed knows how to grow.

Consider the miracle of seeds. Somehow, using nothing but dirt, air, and water, each seed turns into a plant like the one it came from. Explain *that*.

Seeds contain within them a remarkable intelligence. Plants are "born" knowing stuff. How many hours of daylight are needed before they sprout? How warm should it be? How can roots attract the food they need? Which way to grow to absorb more sun? When to start blooming? When to drop seeds so they'll have a shot at sprouting next year? And when to drop leaves, blanketing the soil they're growing in to shelter it for the coming winter? Plants know all this and more.

Seeds come with an internal calendar, timer, thermometer, altimeter, light and moisture sensor, and solar energy locator, not to mention a blueprint for how to make a plant that makes more seeds. And soil knows how to nourish those seeds with exactly what they need, given enough raw materials.

When I have my priorities straight, this fills me with wonder.

TAPPING NEW AND IMPROVED SCIENCE. Growing up in the Bay Area suburbs, amid blocks and blocks of manicured yards, I was well versed in the carpet-bombing, nature-as-adversary approach to gardening. I've since begun to understand the complex system of microbes and organisms already put in place by nature, a system we damage with tilling and chemical applications. My approach eventually devolved to a more primitive one that allows nature to do what it does best: grow stuff.

In my lifetime, science has gone from being a solver of mysteries to a creator of them. When I was young, we naively thought science had pretty much figured out nutrition, for people and for plants. For us, eating well was a short alphabet of vitamins — we built strong bodies 12 ways — and for plants, it was nitrogen, phosphorus, and potassium, and maybe a trace mineral or two. We fortified our white bread with vitamins, and we poured distilled versions of what we thought plants needed onto our gardens, which grew really well. For a while. We had not only unlocked the mysteries of nature, we had improved on it!

Or not . . . Turns out that the concentrated fertilizers that gave plants a visible boost were destroying a system nature had been using for eons — a system dubbed the "soil food web" in *The Soil Biology Primer* by Dr. Elaine R. Ingham, a pioneering microbiologist. A cubic foot of healthy soil contains trillions of living things. *Trillions.* Literally. Bacteria, fungi, protozoa, archaea, arachnids, and worms, all working together to feed plants and, therefore, us.

And that's just what we've discovered so far.

This largely invisible population is driven off or killed outright by the application of garden chemicals. Even organic chemicals, which I use occasionally but less and less, tinker with the intricate soil system of checks and balances that predates people by roughly a bazillion years.

We got a little big for our britches there for a while, we humans, with our ever-evolving science. The fact is, nature doesn't need us to make nutrients in a lab and feed them to plants. Left to its own devices, and given sufficient organic matter (stuff that is currently or formerly alive), the soil food web forms a network for feeding plants that we're only beginning to understand.

Go to a national park, or any place that's wild. Look at a forest or wildflower meadow. Nature has pretty much figured out how to do this without our help, wouldn't you agree?

WORKING WITH NATURE. My approach to gardening is much easier than the one I was raised with, and focuses on giving soil what it needs to nourish the plants I want to grow. Gardening with nature as an ally is simply a matter of offering a variety of foods for germs and worms — the plants will attract the microbes they need to flourish, given enough choices. Diseases are kept in check by the vast population of largely beneficial organisms that "compete or eat," and by plants that, thanks to the soil food web, are robust and healthy.

Gardening in partnership with the soil food web offers many benefits, including:

o   A toxin-free yard and, thus, unpoisoned plants, children, and Chihuahuas
o   More flavorful produce
o   Flowers you can enjoy without fear
o   Healthier bees, butterflies, and other living things, all the way up the food chain to humans
o   Better sleep for the gardener, who is doing right by the planet

There are lots of ways to nurture the soil food web; primarily, you want to shelter your soil by mulching it, adding compost, and by avoiding rototilling, which chops up and dries out the organisms in your soil. These no-till practices provide a safe, moist, and nutritious place for soil organisms to live.

But what many gardeners don't realize is that they can also help sequester (store) carbon by using these practices. The dark color in rich soil comes from humus, the carbon-rich end product of rotted organic matter. Tilling mixes that humus with air, taking carbon out of the soil and mixing it with oxygen to make carbon dioxide. Sheltering your soil keeps carbon in the ground where it belongs, rather than in the air warming up the planet.

In this book, I'll show you how to enjoy your garden more, work less, and harvest more — all while growing rich soil and saving the world, or some portion thereof. You'll learn about the huge, and hugely complex, web of soil life — orchestrated by plant roots — that feeds plants and, in turn, everything else, while pulling carbon out of the atmosphere. And you'll be inspired to nurture that soil life with simple, ongoing practices.

I liken this gardening approach to living a healthy lifestyle, as opposed to struggling with bursts of dieting and exercise. Instead of spending weekends feverishly weeding, tilling, and fertilizing, followed by long stretches of neglect and guilt, you'll learn to enjoy regular puttering that keeps your garden tidy while feeding the soil.

Turn the page for some of my commandments — er, suggestions — for creating healthy, living soil. Think of them as "rules of green thumb," which you'll learn more about throughout the book. Then go take a nap. This is easier than you think.

10 ~~COMMANDM~~ SUGGESTIONS
for Creating Healthy, Living Soil

1 STEP AWAY FROM THE ROTOTILLER. Except in rare circumstances (newly graded or very heavy clay soil), rototilling will only cause your soil microbes sorrow. Tilling destroys fragile but vitally important mycorrhizae (fungal strands; learn about them in chapter 3) and chops up earthworms and other garden friends. Just say no.

2 NO BARE DIRT, NO HOW. Once your soil is loosened, you can do most of your enriching by mulching, which also creates a moist, happy place underneath where worms and germs can frolic and multiply. Mulch well with anything that used to be a plant; this also protects soil from temperature extremes and drying. Dry, sun-baked soil is inhospitable to life.

3 GROW COVER CROPS. Cover crops are basically mulch that's alive. These are plants that shade, feed, loosen, and protect the soil, then die (either on their own or with your help) and turn into soil organic matter. Some, like legumes, even add nitrogen to the soil.

4 YOU DON'T NEED NO STINKING FERTILIZER (MOST OF THE TIME). Feed your dirt with mulch and compost; the plants will take it from there. If you provide a diverse mix of foods for worms and germs in your soil, those critters break down raw materials into absorbable nutrients. Plant roots attract just the right organisms to create a nutritious, pH-compatible root zone. Plants are amazing like that. In the rare cases where fertilizer is necessary (graded or depleted soil), use only organic ones.

5. ADD, DON'T SUBTRACT. When you're tempted to eliminate a bug or disease, first try to add diversity that will "eat or compete with" the offender. Diverse plantings and organic practices will attract good bugs that eat bad bugs. And diverse plantings, along with compost and compost tea, will bring beneficial microbes and bugs and eventually contain most problems.

6. GROW STUFF. Roots feed soil microbes with their sugary exudates; this attracts lots of microbes and other soil creatures that eat each other and die and, along with the roots themselves, turn into nutritious organic matter. Plants improve soil.

7. MAINTENANCE IS MULCH WAITING TO HAPPEN. When weeding and pruning, follow these simple rules: (a) pull stuff up, throw stuff down; (b) snip and flip; (c) chop and drop. Think of this as providing food and shelter for your soil food web. While beautifying and tidying your garden, you're returning nutrients to soil and adding organic matter.

8. DON'T SPRAY THAT WEED! Pull it up and use it to mulch something you love. Weeds are nourishment for the life in your soil, and sometimes for humans. See if it's edible; if so, pour on a little vinaigrette.

9. USE WHAT YOU HAVE. Except in an emergency (your mother-in-law coming to visit for the first time, say), don't haul in truckloads of topsoil and compost; the plant waste growing in your yard or coming from your kitchen will feed the soil. Plant a few things at a time, dig in a handful of compost, and mulch with whatever weeds are growing nearby.

10. EXPERIMENT! HAVE FUN! The best scientists are like little kids — they try new things because they haven't been told what won't work. Give new ideas a shot; either they'll work or you'll find out they don't. If you're not making mistakes, you're not learning anything. And have fun! That, and being nice, are what we're here for, best I can tell.

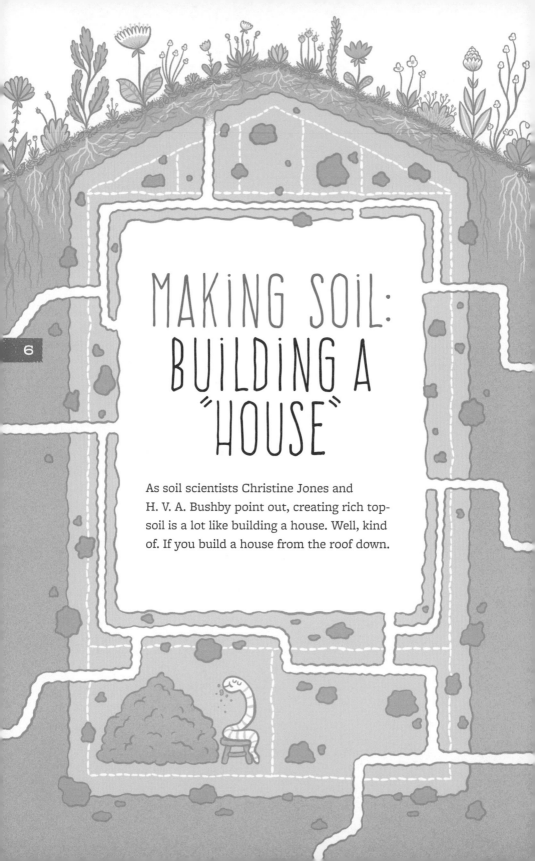

Making Soil: Building a "House"

As soil scientists Christine Jones and H. V. A. Bushby point out, creating rich topsoil is a lot like building a house. Well, kind of. If you build a house from the roof down.

1 START WITH THE ROOF. Soil needs cover to shelter it from the elements: air, temperature extremes, and too much or too little moisture (flooding, erosion, desiccation). Avoid bare dirt; keep it covered with live and/or dead plants — crops, cover crops, or mulch — to protect the soil food web living there.

2 BUILD WALLS. Add organic matter — the stuff that gives soil structure and keeps the house from collapsing. Use sticky microbes to nail (glue) it all together.

3 BE SURE THERE'S GOOD VENTILATION AND PLUMBING. These systems are facilitated by soil organisms that make aggregates or "crumbs," which allow about half of soil volume to be spaces. These spaces are where air and water move in and out.

4 PUT IN A NICE PANTRY. The soil house's "pantry" — the cation exchange capacity (CEC) — creates soil's ability to hold and move nutrients. Clay, organic matter, and biochar are CEC powerhouses; more about them in chapter 4.

5 BE SURE THERE ARE PLENTY OF NUTRIENTS IN THE PANTRY. Plants make the sugars that soil organisms eat, and those organisms make the minerals that feed plants. Biologically active soil feeds itself. If your soil seems dreadful and your plants are languishing, you may need to do a soil test to determine if the right nutrients are available; eventually, though, the soil food web will create a custom blend that's just right.

6 PLANT A LITTLE KITCHEN GARDEN FOR YOUR SOIL FOOD WEB. Use compost and compost tea to feed soil organisms, so they can make carbon-rich humus and adorable little baby organisms.

Once you've built a cozy house, the soil food web will thrive therein; this is what makes soil rich. Plants will make sugars through photosynthesis and put out sweets that bacteria, fungi, nematodes, protozoa, bugs, and worms will come to dine on, creating rich soil for plants — it's a wonderful feedback loop. In the process, they'll constantly be making organic matter and more nutrients.

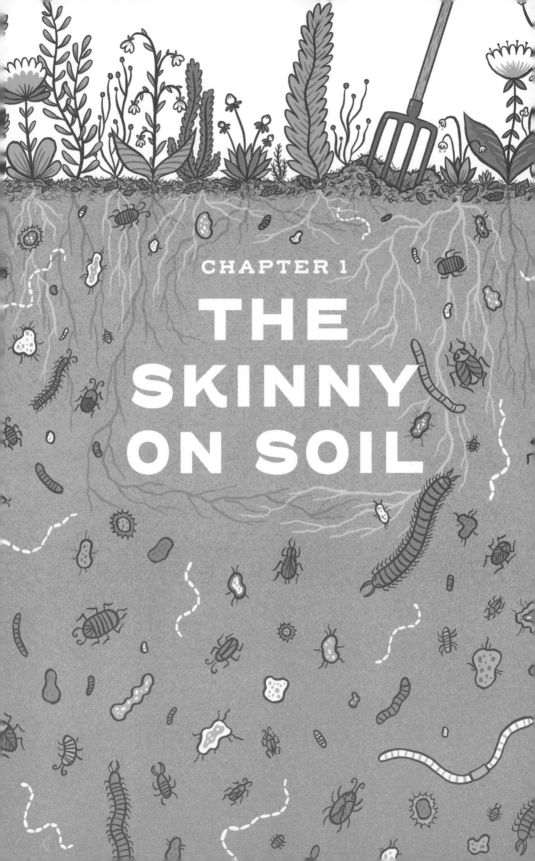

CHAPTER 1
THE SKINNY ON SOIL

What Makes Soil Good & How Does It Get That Way?

Soil is not just something you plant vegetables in and then track onto your carpet. Soil is a living, breathing organism that feeds everything on the planet and also gives us a shot at saving it; it's your favorite planet's third largest carbon sink (something that takes carbon dioxide out of the atmosphere and holds it in the ground).

Every gardener knows that the key to a good garden is good soil. But what is soil? What makes it good? And how does it get that way?

What *Is* Soil?

Soil is composed of roughly:

- 45% minerals (sand, silt, and clay)
- 20 to 30% air
- 20 to 30% water
- 5 to 10% organic matter (anything that is currently or formerly alive, including microbes); this is our favorite part

The inert, mineral parts of soil — sand, silt, and clay — are categorized by size. You know roughly how big grains of sand are. Silt particles are about the thickness of a strand of hair, and clay particles are too small to be seen, until they clump together into giant, cracking clods in your backyard. Clay differs from larger soil particles in that it contains some organic matter and plant nutrients.

Fun facts:

- Soil contains microbes, oxygen, minerals, water, and decaying organic matter.
- Humans contain microbes (90% of the cells in a human body!), oxygen, minerals, water, and decaying organic matter.

Notice a pattern here? I know it's a bit of a blow to the whole "Masters of the Universe" thing, but it's true — we're made of pretty much the same stuff as dirt (usually minus the worms). And our lives depend equally on the lives of microbes. Granted, this mess of minerals and germs is a little more organized in humans — there are clearly marked entrances and exits, for one. But soil as an organism is no less miraculous.

I know it's a bit of a blow to the whole "Masters of the Universe" thing, but it's true — we're made of pretty much the same stuff as dirt (usually minus the worms). And our lives depend equally on the lives of microbes.

Soil is created when nature has its way with the crust of minerals that coats the Earth. Weathering (rain, heat, freeze/thaw cycles) gradually turns rock into sand, silt, and clay particles. This creates something more workable than rocks, but it's living organisms that do the heavy lifting of soil building. As lichen, microbes, plants, and animals move in, they create a cycle of enrichment that gradually and continually makes soil more hospitable to plants, given the right circumstances.

What Makes Good Soil Good?

The first thing you probably notice about soil is its texture, which is determined by the proportions of soil particles — sand, silt, and clay. This proportion in any given area isn't easily changed.

- **Sandy soil** is gritty. It drains well but tends to dry out. It's not good at holding on to water and nutrients.
- **Silty soil** has a softer texture and holds more nutrients but washes away easily.
- **Clay soil** is slippery, gummy, or rock hard, depending on how wet or dry it is. It holds nutrients well but compacts easily and doesn't allow much water or air to pass through. Clay carries the most clout — a small amount of clay has a strong influence on soil texture.

Good soil contains roughly 40 percent sand, 20 percent silt, and 40 percent clay; this mix has a crumbly texture that holds water and air and drains well. You are more or less stuck with the soil you've inherited, in terms of particle proportions, but you can improve the texture by adding compost and other organic matter such as sawdust or fine wood chips. The soil food web then breaks down this organic matter and glues it to soil particles to make crumbs, or aggregates.

Gypsum has a reputation for breaking up clay soils, but the research on this is varied, and the effect is temporary — maybe a few months. Organic matter is longer-lasting and adds life, as opposed to just loosening up clay.

PERLITE, VERMICULITE, AND ASBESTOS

There are two other things you can add to improve soil texture: perlite and vermiculite. Both are rock products that are very absorbent, and thus stabilize soil moisture. Their size — similar to coarse sand or fine gravel — also adds variety to soil particles and loosens clay soil. You'll notice light-colored pieces of perlite or vermiculite in most commercial potting soil.

Perlite descends from a volcanic glass, usually obsidian. Vermiculite is a type of rock composed of shiny flakes resembling mica. Both are mined and then processed by superheating, which causes them to puff up like popcorn and become absorbent.

One of the early sources of vermiculite was the Libby Mine in Montana, whose vermiculite turned out to be contaminated with asbestos dust (this was before we knew the dangers of asbestos). There's ongoing concern that vermiculite contains asbestos, but it no longer does. The Libby Mine closed in 1990, and no mines currently in use contain asbestos.

Really good soil, the stuff organic gardeners drool over, contains a *lot* of life. Sure, there are structural components (soil particles) and minerals (especially N-P-K, which stands for nitrogen, phosphorus, and potassium — the letters on the chemical fertilizers you won't be buying) that affect the quality of soil. But simply combining the structural components with minerals will give you a mixture that neither contains nor supports life.

To deserve the name "soil," this mix must contain life. A teaspoon of good soil contains more microbes than there are people in the United States. Plus, good soil holds organic matter and arthropods, worms, and other crawly things that keep the whole wonderful mess alive, crumbly, and nourishing to plant life.

 A teaspoon of good soil contains more microbes than there are people in the United States.

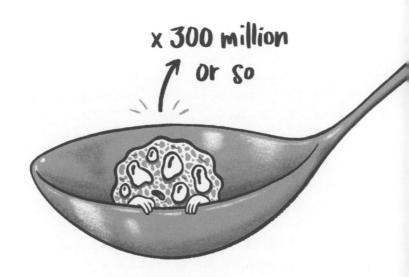

x 300 million or so

Organic Matter

Soil organic matter (SOM, to friends) is probably the most important part of good soil. Organic matter consists of living or formerly living things, in various stages of decomposition. We'll talk more about SOM in chapter 3.

High-quality topsoil is created largely by plants pulling carbon out of the air via photosynthesis, turning that carbon into sugars that feed soil organisms in their root zones, and then turning their dead plant selves into organic matter. One end product of all this digestion is humus, a stable form of soil carbon. And all that activity anchors carbon in the soil instead of releasing it into the atmosphere.

It can take eons of natural forces to produce high-quality soil. Or, given the right raw materials and the help of, say, a gardener like you, it can happen in a matter of months.

What Good Is Good Soil?

So what's the big deal about having good soil? Can't we just plunk things in the ground and fertilize them?

Well, kind of. The problem is, relying on manufactured fertilizer means you'll be growing inferior produce, depleting your soil, and causing grief to the planet.

Healthy, living soil pulls carbon from the atmosphere, soaks up water to prevent floods, is less vulnerable to drying, and feeds plants exactly what they need while controlling diseases and pests. It becomes richer and richer because of all the things that live and die in it. In effect, healthy soil is an ongoing composting process.

Soil that is fed chemical fertilizers, on the other hand, "goes dead." The organisms that would be living there and feeding plants are chased off or killed by concentrated fertilizers, which give plants a blast of a few nutrients but not the sustained, customized mix of minerals and trace nutrients that the soil food web can offer. It's like feeding kids vitamins instead of fruits and vegetables — you've given them a shot of something that's a small part of what they really need. Instead of rich, composting soil, you're left with just the mineral components — sand, silt, and clay — and some chemicals. What's missing? Life.

Chemical fertilizers also render soil organisms that feed plants obsolete; the fertilizers do the job in a quick and soulless way, kind of like self-checkout aisles at the grocery store. The checkout clerks, and the soil organisms, go sadly on their way to sign up for unemployment or eat out of Dumpsters.

Providing rich, biologically active soil is the best thing you can do for your plants. The soil food web arranges to bring exactly what any given plant needs right to its rhizosphere, or root zone — sort of a Domino's pizza for plants. If Domino's delivered perfectly nutritious health food.

This soil food web process creates an area of ultrarich soil in the rhizosphere; this happens because bacteria and other organisms that eat root exudates then die and feed other members of the soil food web in the neighborhood. All those dead organisms are organic matter — nutrient-rich, sticky, crumbly bits of wonderfulness. After a certain number of trips through the digestive tracts of soil food web members, the carbon in organic matter becomes humus — a long carbon chain that provides a home for water, air, and nutrients.

The root zone is part of a feedback cycle where the roots feed microbes, microbes feed the soil, and the soil feeds more roots. Pull a little clump of grass out of some gravel by the roadside — you'll see dark, rich soil clinging to the roots. The soil food web did that.

Good soil, then, grows good vegetables and flowers and trees and shrubberies, and begets more good soil. It also sequesters carbon; your planet appreciates this.

Chemical fertilizers also render soil organisms that feed plants obsolete; the fertilizers do the job in a quick and soulless way, kind of like self-checkout aisles at the grocery store.

Soil, Climate Change, and Saving My Grandchildren

Perhaps you've heard of global warming. And perhaps you know that carbon dioxide is one of its primary causes. What does this have to do with soil, you ask? Plenty. Read on.

Soil is not just a place to stick your seedlings. Soil is this planet's third largest carbon sink (reservoir that holds carbon), after oceans and fossil fuels. It's a huge, carbon-sequestering, water-conserving, dead-stuff-recycling, life-sustaining organism that covers around 30 percent of the Earth.

Healthy, live soil is constantly pulling carbon dioxide from the atmosphere and sequestering it — turning it into organic matter and humus, the dark stuff that's the hallmark of good soil. Soil affects climate change. And, happily, the best soil for your planet is also the best soil for your plants.

Captain Soil gobbling up carbon dioxide and thereby saving planet Earth, where my grandchildren happen to live. I love Captain Soil.

CARBON STORES ON EARTH
(IN BILLIONS OF METRIC TONS)

Soils sequester (store) more carbon than the atmosphere and all plant life combined.

Oceans: 40,000
Fossil fuels: 4,000
Soil organic matter: 1,500
Atmosphere: 800
Plants: 600

Sadly, humans are especially good at putting that carbon back into the atmosphere, by burning fossil fuels, of course, but also with our agricultural practices. Those beautifully plowed or rototilled rows we romanticize are actually big cemeteries for mycorrhizae (my-kuh-RY-zee; literally "fungus roots"), worms, and germs — all the stuff that makes soil a living organism. And they're kind of a speed-dating site where soil carbon molecules meet atmospheric oxygen molecules. Once introduced, carbon and oxygen marry, in a bigamous sort of way (one carbon molecule to two oxygen molecules, hence the name CO_2), and float off together into the heavens as carbon dioxide.

This marriage is ill advised, for a couple of reasons:

o **Our soil needs carbon.** Carbon is the dark stuff that is contained in humus; humus is what makes good soil dark and good.

o **Our atmosphere doesn't need more carbon dioxide.** CO_2 is acting like a gigantic floating row cover over the Earth, making the planet WAAAY too warm. Global warming was a good thing in the past, during, say, the Ice Age. Now, however, we've become accustomed to ambient temperatures below 120°F (49°C). I prefer it this way.

There are a couple of ways that gardeners can help keep carbon where we need it: in the soil. We can decrease the amount of carbon that is released into the atmosphere (the supply side) and increase the amount of carbon that is trapped in the soil (the demand side).

TO REDUCE THE SUPPLY, we can stop tilling. Tilling systematically degrades soil, essentially "burning" (oxidizing) carbon that's living in it, often after centuries of topsoil building on the part of Ma Nature. Tilling in plant residues adds some carbon back, but this green matter gets burned through by bacteria, oxidizing it and making more CO_2; the overall effect is a net loss of soil carbon. Cultivated soils lose *50 to 70 percent* of their carbon stores.

TO INCREASE THE "DEMAND" FOR, or use of, carbon, we can foster the growth of things that use it — plants and soil microbes being foremost.

Sinking carbon into soil and keeping it there has far-reaching benefits. Soil that's rich in organic matter and, therefore, carbon, is more fertile, has better texture, and is able to buffer pH. This rich soil also acts like a sponge; it holds moisture for when plants need it and absorbs water from heavy rains, thereby helping prevent floods.

Soil that's degraded by destructive agricultural practices such as tilling and chemical applications, on the other hand, is unable to store water or nutrients. Soil quality drops, "requiring" more water and chemical fertilizers and pesticides that then run off into, say, Lake Tahoe and turn it green.

 Soil that's rich in organic matter and, therefore, carbon, is more fertile, has better texture, and is able to buffer pH.

WHY YOU CARE ABOUT SOIL CARBON, IN A NUTSHELL

- Increased soil carbon leads to increased soil life, which leads to increased soil fertility, which leads to better storage of water and nutrients and better (crumblier) soil texture.

- Carbon that's in soil is carbon that isn't in the atmosphere. Sinking carbon into soil helps save the planet.

- All of this leads to better tomatoes.

Franklin D. Roosevelt wrote in 1937, "A nation that destroys its soil destroys itself." The same could be said about the planet that destroys its soil. Each gardener or farmer who stops tilling and starts building soil life not only improves their soil but also pulls a little more carbon out of the atmosphere.

The future of soil depends on us. And the future of humans depends on soil. Really. I'm not even exaggerating.

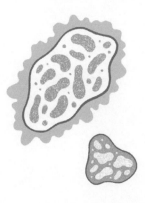

Start with the Roof (Mulch & Cover Crops)

Why start with the roof? Because your soil food web family needs shelter from the elements. Plus, whatever's on top of your soil will become humus, once the soil food web has its way with it. And great soil is full of humus.

Different types of mulch have different end results; trees and shrubs contain more lignin — the stuff in woody plants that makes long-lasting humus — and green, annual plants contain more cellulose, which tends to be digested into plant food. Both, though, contribute to making soil food and, eventually, humus. Mulch can be dead or alive: cover crops you plant, existing weeds, garden waste, or any other organic matter you lay on the ground.

Mulch

Mulch does many things for your soil. First, it shades and softens it, and dramatically decreases the amount of watering required; your soil will be moist and easy to "spot-dig" when it comes time to plant. More important, though, mulch can provide everything needed for nature to start doing its magic, via the soil food web (more about that in chapter 3). Healthy soil is alive soil; that life needs moisture and organic matter to thrive.

Mulch can be made from lots of things: cardboard, thick sheaves of wet newspaper (wet because you want it to stay on your soil, not, say, in your neighbor's yard), straw or hay, leaves and other garden waste, or the plants you just pulled up. The more — and more diverse — your mulch, the better. There just has to be enough of it to prevent sunlight from reaching the plants you're smothering.

Mulch is magic! After a month or so under a deep layer of organic matter that has been kept moist, your garden soil will be soft and full of worms. When that happens, you've basically turned your entire planting bed into compost.

WHAT NOT TO MULCH

Don't throw long pieces of anything onto your garden mulch (or into your compost, with the exception of your long-term rough compost). You're making a salad bar, not a brush pile. Also don't mulch with plants that are invasive or hateful, like Bermuda grass or poison ivy.

MULCH: THE GREEN & THE BROWN

The soil food web loves green, juicy organic matter. It also adores dry brown matter, but that's a less passionate love; it takes a while for brown matter to consummate to become humus, and few nutrients are produced in the process. Brown matter tends to create stable humus, while green matter encourages a bacterial feeding frenzy and adds more plant-available nutrients to the soil. So you want both.

Brown matter attracts more fungi than bacteria, leading to a slow but steady decomposition and a lower pH. Green matter attracts bacteria, which leads to livelier, more nitrogen-rich soil and a pH in the ideal range for annual garden plants like tomatoes. In general you'll want about one-third green to two-thirds brown matter (just like when you're making compost).

Whatever's on top of your mulch will end up being brown matter, since it'll be exposed to the drying effects of sun and air. But why let juicy green mulch — cover crops, weeds, or spent plants you pull up and lay down — turn into brown stuff? Instead, place a layer of brown matter, such as straw or leaves, over a layer of fresh green mulch.

Interesting side note: When deciding what to use for mulch, match the mulch to the plant. Green, annual plants like more green in their mulch, and they like it finely chopped to attract more bacteria, which make more nitrogen and raise soil pH. Bacteria like to binge.

Browner, woodier plants prefer browner, woodier mulch. Trees and shrubs thrive in fungally dominated soil, and brown material is what fungi love. Fungi are slow-but-steady diners.

Cardboard

Cardboard is best used for landscaping untamed areas where weeds are hard to control (think blackberries, Bermuda grass, dock, and shotweed). It's also the best way to turn lawn into some other kind of garden landscape; this new landscape can range from a bed of bark mulch and drought-resistant shrubs to a new vegetable garden. Cardboard kills all the plants underneath it by starving them of light; you can then plant the things you want there while the cardboard slowly breaks down.

Start scrounging for cardboard any time. Appliance stores are great places to get giant pieces, which are the easiest to use, but any decent-sized boxes will do. You can often find boxes behind grocery stores, usually already flattened.

With cardboard in hand, begin by cutting everything growing in your selected location *very* close to the ground. Leave the cut plant material in place to compost. After a good rain or thorough watering, lay a single layer of cardboard (overlapping at the edges) over the area. Cover any gaps that expose soil with a whole section of a newspaper, soaked in a bucket of water. Extend the cardboard over the edges of the planting bed, if possible, to keep weeds from creeping back in, or pull up plants around the edges and tuck them under the cardboard. Fall is a good time to lay down cardboard, as autumn and winter rain will keep it moist enough to cling to the ground and rot.

Water down the whole thing again, then sit back and watch the cardboard work its magic. This will be quite dull — you might want to bring a book. Alternatively, find something else to do while the cardboard works: go about your daily life for a season or so, knit a sweater, then come back, lift the cardboard, and marvel at the composting goodness underneath.

Mulching with cardboard works best for areas you can mostly leave alone for a while — at least fall through spring. Cardboard will take about a year to break down, less time if you keep it moist and covered with mulch (which looks nicer) and longer if you ignore it. I used cardboard around a fence where Bermuda grass was thriving; it killed the grass in 6 months, and the area became respectable looking once I raked mulch over it and planted climbing roses and raspberries.

You can plant small areas immediately in the cardboard mulch. This works well for shrubs and seedlings, not so much for seeds. Cut an X in the cardboard with a butcher knife, fold back the points, chop in some compost, and plant. Fold the X back around the base of the plant to shade the ground, then cover the whole bed with bark mulch. Be on guard for weeds rising through the gaps; the more holes you cut in the cardboard, the more places there are for unwanted things to grow through. I prefer to plant just a few shrubs in a new cardboard cover — just enough that the area appears loved. A few shrubs in a wide expanse of bark looks surprisingly finished.

You can also plant veggies immediately, but use seedlings (not seeds) and be sure to toss compost and/or a natural, nitrogen-rich fertilizer in each planting hole. When it's time to pull up the veggies, also yank any weeds that sprouted around the plants. Cover the hole in the cardboard with a section of wet newspaper, lay the spent plant and weeds down as mulch, and make holes in different parts of the cardboard for what you plant next. Over a year, this process will kill even noxious weeds and turn them into compost. Keeping the area mulched with straw will help break down the cardboard so it, too, eventually becomes part of the soil.

Cardboard kills lawn or weeds, which will now turn into compost. Chop holes in the cardboard, plant shrubs or trees, mulch with bark, and voilà! You now have landscaping!

Straw and Hay

Straw and hay are not the same thing. Hay is food for livestock — horses, cows, and goats, for instance. It contains nutritious leaves *and seeds*, which may well sprout in your garden and cause you grief. Straw, used primarily for animal bedding, consists of stalks left after the edible grains have been removed. It's mostly made of wheat or rice, and it doesn't drop seeds.

Both straw and hay function as mulch, shading and feeding the soil food web. Alfalfa hay contains the most nitrogen, feeding faster-growing green plants and bacteria. If you're starting with barren soil, alfalfa hay is a good choice. It contains few seeds, so it saves you the aggravation of having planted new weeds. If you mulch deeply enough, though, seeds won't have enough light to germinate and grow.

Ruth Stout, author of *The Ruth Stout No-Work Garden Book*, introduced me to gardening with mulch while I was killing time at the Cal State Library turnstile in my first job, back in 1971. This book (despite its somewhat misleading title — I'd call it *The Ruth Stout Less-Work Garden*) made that time-killing worthwhile. Stout recommended mulching with a very thick layer of straw or hay — about 12 inches — to shade out unwanted growth and create a rich layer of compost. Hers was a revolutionary approach at that time and is now an important aspect of permaculture gardening.

Other Materials

Anything that grows in your garden — including weeds and spent plants — can become mulch. When your annual veggie plants are done serving their purpose, for instance, pull them up, chop them up, and throw them down on the soil as mulch. If the whole bed is finished, pull and chop up everything, and rake brown duff from the surrounding area over the whole thing; this works out well, as most plants are "done" in fall, when there are lots of leaves, pine needles, and dried-up garden plants to serve as a brown layer.

Everything produced in your garden can be used. Smaller garden waste (weeds, soft prunings, chopped up spent plants) can be added to a compost pile or sheet composted. Larger branches can be chipped, or burned to make a form of charcoal known as biochar (see chapter 4 for how). These practices set up a regenerative cycle in your garden, where you harvest both produce and soil nutrients, while producing no waste and continually enriching your soil. Everything that grows goes into building your soil house.

Cover Crops

As the old saying goes, "nature abhors a vacuum." (Nature's a lot like me that way; check out my living room rug if you doubt it.) If you leave dirt uncovered, nature is likely to put some weeds in it. One way to avoid a vacuum — the bare dirt kind — is to plant cover crops (basically living mulch).

Sometimes called green manure because they add nitrogen to the soil, cover crops accomplish several things. By sending roots deep into the ground, they break up soil clods into crumbs, insert organic matter, attract soil organisms, and, when they rot, create empty channels of various sizes for the soil food web to live in. The green plants, once you cut or flatten them and mulch over them, decompose into nitrogen-rich compost while feeding soil microbes.

Cover crops come in roughly three categories:

o **Grasses.** These create lots of biomass (organic matter) but can be messy looking and hard to eradicate in established beds.

o **Broadleaf plants.** These shade soil, coexist nicely with other things you've planted, look presentable among the things you're growing, and are generally easy to manage.

o **Legumes.** These fix nitrogen — that is, pull it out of the air and anchor it in the soil for plant food. You may be surprised at how many legumes there are, and how many of them do well in a soup pot.

Many edible and ornamental plants can act as cover crops by spreading and shading the soil or by generating bodacious amounts of biomass. More about those shortly.

News flash: Plants make soil. In order to survive, all plants enrich and add to the soil in their root zones; they do this by convincing the soil food web to bring them nutrients in exchange for sugary root exudates. If you pull a weed or clump of grass out of sandy soil, you'll see rich, dark dirt around the roots. This is more than coincidence. That plant didn't happen to find a hunk of root-shaped dirt to grow in; no — it *created* it by attracting organisms that then decomposed into soil. This is how the progression of soil begins: plant roots create soil, which grows more plants, which rot and create more and richer soil.

If you pull a weed or clump of grass out of sandy soil, you'll see rich, dark dirt around the roots. This is more than coincidence. That plant didn't happen to find a hunk of root-shaped dirt to grow in; no — it *created* it.

Cover crops also compete with weeds. You can pretty much eliminate most types of weeds in a planting area by generously seeding with a cover crop (pull up weeds, shake off the dirt, and lay them down over the seeds), then pulling everything up or cutting it down before any new weeds go to seed.

My Favorite Cover Crops

So we now know the power of cover crops, which will loosen and enrich the soil in their root zones while alive, then rot into beneficial compost. The following are some of my favorites, which I might plant singly or in combination, depending on what I'm shooting for.

BUCKWHEAT

Buckwheat is an easy, fast-growing cover crop. You can toss and scratch in seeds any time, onto any unoccupied space in the vegetable garden. Because it's so quick to germinate and grow, buckwheat will provide soil organic matter and green mulch by the time you're ready to plant the next thing. If it's allowed to flower, buckwheat also feeds pollinators — including honeybees — and reseeds itself; that's a good thing with such an easy-to-manage cover crop.

When it's time to plant that area again, and when the soil is moist, just cut or pull up the buckwheat and lay it down on the planting bed. The roots become soil organic matter if you leave them in the ground, or they soften the soil as you pull them out, leaving behind some root hairs to rot. The green plant tops on the surface both shade the soil and feed it as they decompose. Move the new mulch aside, chopping holes or rows, to plant your next crop.

SOYBEANS AND OTHER, LESSER BEANS

I love soybeans as a cover crop for two reasons: (1) they fix nitrogen, and (2) they make edamame.

All legumes fix nitrogen: they pull it out of the air and stick it into root nodules that then feed whatever plants are growing in the neighborhood. You can see these nodules when you pull up bean plants; they look like little pink warty things on the roots. Because legume plants are bigger and slower growing than other cover crops, use them in areas you won't be planting for a while — usually areas that will lie fallow for a season or new areas you want to cultivate later.

Bush peas or green beans also work as a cover crop. The challenge with all edible cover crops is to remember to hunt them down and eat them — it's easy to lose bean and pea pods in the revelry of a mixed cover crop. It's fun, however, to grow your own edamame. All the pods tend to ripen at the same time, so you need to make only one or two forays to pick them and pop them in the freezer (no freezer prep required — when you're ready to eat them, just drop a handful in boiling water for a minute or so, drain, salt, and eat joyfully, because they're delicious). Keep an eye on soybeans when they begin to form pods. It's a short hop from edamame to dry soybeans.

For other beans with a longer harvest stretch, you can pick and eat them on the spot as they ripen and as you notice them. If you find any legumes past their prime, let them ripen into dry beans and cook them in soup. Any pods you miss will drop seeds and sow more cover crops. It's a win-win-win-win situation.

Fava beans are in a category of their own. They grow anywhere and produce copious amounts of foliage while fixing nitrogen with their deep roots. The young foliage and the beans themselves are edible and delicious; they're one of my favorite cover crops in uncultivated soil I plan to plant soon.

OILSEED RADISHES

Also known as daikon or forage radishes, these are plants that germinate with gusto and grow thumb-sized roots that break up soil to a depth of up to 6 inches, then leave open channels and organic matter behind. Plant seeds in fallow or uncultivated areas. Be aware that over time the radishes may pop up throughout the garden, which is not a bad thing. If they're encroaching on a planted area, yank them out and eat them. They are easy to pull, which makes them a nonnuisance. You can leave them to run their course, making bigger roots and reseeding themselves, or pull them out and lay them down if you're ready to plant something there. And you can eat the radishes and sauté the greens in oil and garlic. Be sure to strip the leaves off the woodier stems.

OATS AND ANNUAL RYEGRASS

Oats and ryegrass are good nitrogen builders and holders, and good producers of biomass, but they can become weeds in areas with mild winters. If you get hard, reliable freezes, you can count on that cold to kill off annual grasses. But if you live where it *may or may not* freeze, your oats and ryegrass *may or may not* be a blessing. They'll also drop seeds if you don't pull or cut them before they mature. Use them in uncultivated areas where you plan to yank everything up and lay it down, then cover with a light-excluding mulch. Be sure to do this before the plants go to seed and get away from you. Grasses as weeds are messy looking and can be hard to control once established.

MIXED COVER CROPS

In larger areas that will be fallow for a while, or in spots that haven't been cultivated previously, you can plant a mixture of cover crops. Sowing different plants provides a bigger variety of nutrients and attracts more diverse microbes. Make your own mix or buy one.

Scatter larger seeds first (legumes, mainly — fava, soybeans, cowpeas), then pull up everything growing there and shake off the cover crop seeds along with the dirt from the existing roots; this mixes the big seeds into a nice bed of soil.

Lay down everything you've just pulled up to compost in place. Next, scatter buckwheat and daikon radish seeds and grasses (if you want to take that particular leap), add a shallow layer of brown mulch (dead leaves, straw or hay, dried-up weeds) to keep the soil moist, and let it all grow until it's time to plant again. You can do this in fall so the cover crop will be watered by winter storms, then pull it up in spring when you're ready to plant. You can, of course, plant the legumes in spring and harvest them. You can even interplant other crops, like corn and tomatoes; the legumes will feed them nitrogen and phosphorus. Just pull up chunks of the cover crop and shake off the rich dirt where you want to plant.

Covering the area with brown mulch adds to the composting action by keeping the material under it moist, and that brown mulch will eventually turn into compost as well, with the mix of greens and browns that forms good compost (see more in chapter 6).

SWEET POTATOES

This is a brilliant idea I got from Anna Hess, author of *Homegrown Humus*. Sweet potatoes are easy to grow from slips. You make these by sticking toothpicks in a sweet potato and submerging the lower half or so in a glass of water, kind of like what you do with an avocado pit. When sprouts develop, cut them off and root them in water, then plant them in potting soil and baby them until you're ready to plant them in an irrigated spot.

Sweet potatoes produce massive amounts of foliage. Plant them in an area where you want lots of leaf cover, such as with tall plants like corn and tomatoes, or in as-yet-uncultivated parts of your yard. They shade the soil and starve weed seedlings, then provide you and your soil with food.

Sweet potatoes are also delicious. You can pick young leaves and eat them like any greens, ideally slathered with butter (they're not poisonous like regular potato leaves). Then in fall, dig up the tubers to make the sweet potato pie you're hankering for, and use the copious amounts of biomass for mulch.

Chop up the leaves with your machete (see chapter 8 for the how-to). Or, to use a string trimmer as the chopper, lay the leaves down and hold the trimmer at an angle to chop them. Rake the spattered debris back onto the planting beds. Alternatively, you could cut the foliage with clippers into pieces about 4 inches long, but this is labor intensive. Get into a meditative state before you start.

This "chop and drop" process is best done when it's about to rain, so that the green matter you drop on the ground will start to rot instead of dry and become brown matter. Cover the green mulch with brown mulch to help it rot; you might rake leaf litter from trees onto the area to protect the new green mulch from drying out before it rots.

Sweet potatoes starve weeds while feeding the soil.
Plus, they feed you. They're cool like that.

COVER CROPS: WHAT TO PLANT, WHEN

All cover crops provide biomass (organic matter), as both roots and mulch. And all feed the soil food web, making more nutrients and making your soil more bioactive (alive). Legumes such as beans, peas, vetch, and clover also fix nitrogen, pulling it out of the air and sticking it into the soil where other plants can use it.

I prefer to plant a mixture of cover crops, usually a mix of legumes and broadleafs like buckwheat or oilseed radish. I do this because (1) they attract more varied pollinators; (2) some cover crops attract destructive bugs and worms, and a mix will be less appealing and also attract predators; and (3) I love diversity. A mix of plants invites a mix of soil food web inhabitants, making soil richer and more alive.

In areas that are fallow, where you can smother the whole mess, you can grow grasses as a cover crop. Don't use them in areas you're cultivating; they look messy and are hard to eradicate.

There's a lot of leeway in when you can plant cover crops. The main caveat: with the exceptions noted on the following page, don't plant them in the dead of winter — most seeds will just sit and rot.

PLANT ANY TIME

Buckwheat · Crimson clover · Fava beans · Oilseed radishes

The above plants seem to do just fine napping all winter and sprouting when conditions are right in spring. They'll also sprout in summer and fall, and all of them coexist nicely with your cultivated plants, like veggies and flowers. They're easy to pull out if they get too big for your purposes, and by then they will have created a nice root system, breaking up deeper soil and adding organic matter.

Be aware that clover can become invasive if you let it go to seed. Prevent this by mowing it or by flattening and then covering it with a light-excluding mulch. The others can also reseed, but they are easier to yank up or kill by mowing. Avoid planting invasive things in established beds; it's better to reserve them for new territory where you'll be killing off the whole cover crop.

FALL PLANTING

Most grasses · Fava beans · Field peas · Vetch · Clover

Fall is the most common time people plant cover crops. This is because fall and winter rains help seeds germinate and grow, and many planting areas are idle during the winter anyway. Some of these plants are killed by freezes, turning them from cover crop to mulch. Others need to be mowed or smothered.

SPRING & SUMMER PLANTING

Sudan grass (a deep-rooted grass that adds lots of biomass to the soil) · Soybeans · Cowpeas · Mustard

FLOWERS & EDIBLES
that work as cover crops

Sweet potatoes · Strawberries · Sunflowers · Beans & peas · Poppies · Nasturtiums

Plant these in the warmer temperatures of spring and summer, just like most vegetables and flowers.

Turning Cover Crops into Mulch

So, your beautiful cover crop is thriving, but you'd like to grow something else there now. How do you get to the part where cover crop turns into compost? First, you need the existing plants to be in complete contact with the ground so they'll rot. Otherwise, they may grow back before they decompose, or dry up in the air space left. There are several ways to get a cover crop grounded, so to speak: pull it up, cut it down, or flatten and smother it.

PULL IT UP, THROW IT DOWN

The simplest way to start composting cover crops is to pull them up and lay them down. Do this when the soil is moist, after either a good rain or a good watering. Yank everything out, shake off the rich dirt, and lay the green plants on the surface. Then cover them with something brown like straw or dry leaves you've raked up. Voilà! You have greens and browns, the ingredients of compost. This process also loosens the soil where the roots had been growing and leaves behind

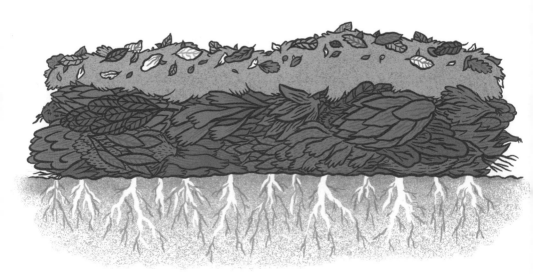

Pulled-up and thrown-down cover crops, topped with brown mulch, will become compost. Note the new channels in the soil where there once were roots.

organic matter in the form of root hairs. The soil stays shaded and protected and the soil food web is never massacred. This is gardening as soil building.

The pull it/throw it approach works whether you're starting a fresh bed or working around perennials or flower and vegetable plants that aren't finished for the season.

For a new bed of annual veggies, pull up and chop the cover crop and weeds (which act as cover crops), spread the chopped material randomly, and cover the whole area with brown mulch. To plant this new bed immediately, scoot the green and brown mulch a bit away from the planting hole, as the decomposing mulch will tie up nitrogen for a while, then add compost or nitrogen-rich fertilizer (manure, alfalfa meal or pellets, blood meal, or feather meal) to each planting spot. This will ensure your new plants don't go hungry while the compost happens around them; before long that compost will be feeding the soil. In some parts of the country, you'll want to scatter organic snail bait, as young plants and seedlings are especially vulnerable to death by mollusk.

For a bed where annuals (vegetables and flowers) are used as edible cover crops among perennials (such as blueberry bushes, strawberries, and asparagus), you won't pull all these cover crops out at once. Instead, remove the spent plants (kale and cauliflower, for instance) then chop up and lay down anything you don't eat and plant the next things, which might be peas, spinach, radishes, lettuce, and new kale among the onions, strawberries, and borage that are still thriving. You might add nasturtium and alyssum seeds for beauty and diversity. When any plant runs its course, pull it up and plant something new there — whatever is in season — adding a handful of compost to each planting spot.

I use the pull it up/throw it down technique throughout the growing season in my curbside garden. What I've pulled up is applied as mulch around existing plants — this includes both cover crops and weeds that act as cover crops. When loose duff is raked over the green mulch, the final product looks tidy, in a natural sort of way.

This method has limitations. Plants with tenacious or matted roots — Bermuda grass, blackberries, dock, and yarrow, for instance — are hard to pull up. These have to be cut or flattened, starved of light, and kept moist enough to rot. Read on.

CUT IT DOWN

Depending on how large an area you're dealing with, you can cut plants to the ground with hedge trimmers or a string trimmer — just don't walk where you plant. You'll want to make paths from which you can reach all your garden beds without stepping on the soil (more about this in chapter 4). Stand on these paths when you cut, pull, or whack your cover crop. Leave the clippings on the beds.

Planting into cut-down cover crops can be tricky. Some plants, especially grasses, will want to keep growing, and you end up with a weedy mess of a planting bed. I find it works best to cover the bed with a light-excluding mulch — either cardboard or 6 inches of straw or hay — and make holes in that for planting. Or cover and wait a season until the cover crop starves from lack of light.

FLATTEN AND SMOTHER IT

You can also just flatten the weeds or cover crop if they're tender enough to stay where they're put, then cover them with enough mulch to keep sunlight from reaching the plants (this could be cardboard, newspaper, or a 4- to 6-inch layer of straw). Sun is what lets plants produce food for themselves; if you exclude sun for long enough, the plant and its roots will die and compost. This may not work for sturdy grasses or other plants that don't like to bend, but it's worth a try. If you find that your mulch or cardboard is making little tents, you'll have to cut or pull up the offending tent poles.

Smothering works best on an area you can leave alone for a while — new ground or somewhere you don't plan to plant for a season. The thick mulch layers are essential; you want all that plant matter to rot into the soil, not become a crop of weeds or dry up into something less interesting to the soil food web.

The benefit of smothering and composting in place is that the roots stay in the ground and rot, adding organic matter deep in the soil and creating food and shelter for the soil food web. Plus, it's easier. You don't have to lug plants to the compost pile and then lug compost to the garden. You can instead, say, take a nap.

EDIBLE & FLOWERING GROUND COVERS

Strawberries and nasturtiums are a couple of my favorite edible ground covers. When my strawberries have too many runners, I dig some out on a rainy day and tuck them into other parts of the garden. They're beautiful, lush, and delicious. Nasturtiums make abundant seeds; just pull off the seed heads and tuck them under your mulch to grow more. The flowers add beauty and a delicious radishlike flavor to salads. You can also pickle young nasturtium seeds to use like capers.

Sweet potatoes produce generous amounts of (edible) greens, as well as . . . sweet potatoes.

Alyssum, California poppies, and spinach or salad greens make a gorgeous and partly edible ground cover mix. And climbing peas, if grown without tall supports, will clamber over the ground and function as ground cover; you can also eat the young shoots, sautéed in olive oil and garlic.

Remember, plants improve the soil. A good, diverse ground cover under your vegetables feeds the soil food web, bees, and beneficial bugs, and it keeps the soil moist. If that ground cover is in the way of something you want to plant, just pull up a section, shake off the dirt, and use the plants to mulch your new ones.

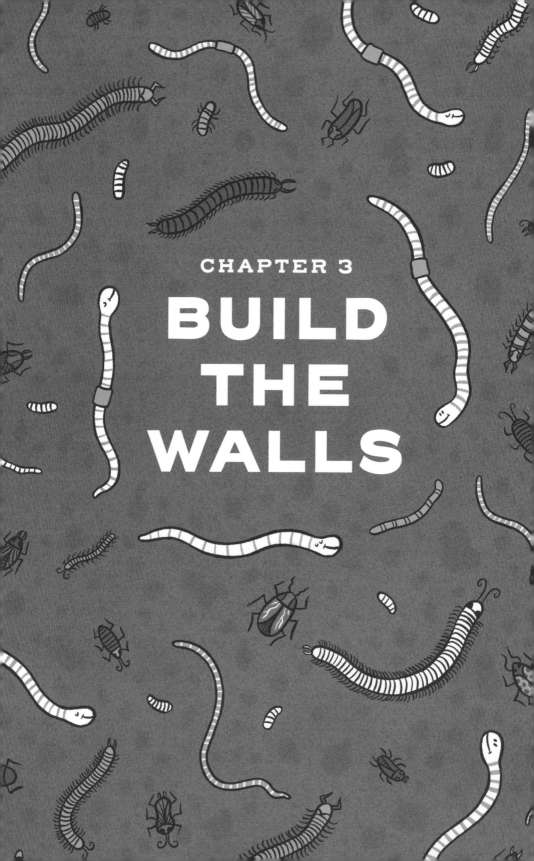

Organic Matter & the Soil Food Web That Sticks It Together

Organic matter is a key part of healthy soil. It feeds the soil food web, which feeds plants; all of those contribute to the "walls" of our soil house. Organic matter in various stages of decomposition forms the building blocks of your house; it creates soil particles in different sizes and, eventually, humus. Members of the soil food web, in the process of feeding on organic matter and each other, add still more variety to the size of particles (think tiny dead bugs). And certain members of the soil food web — worms — create some of the most fertile and best-textured soil around, in the form of worm poop or, to put it more delicately, castings.

Soil Organic Matter

Soil organic matter (SOM) makes soil a self-correcting entity. SOM holds water, improves air circulation, and nurtures the diverse microbes that feed plants, buffer pH, and keep pathogens in check. Organic matter also provides the materials for soil's "building blocks" — aggregates of mineral particles, germs, worm castings, and bits of dead bugs. These crumbs are what gives good soil its structure.

About half of SOM is made up of carbon — that stuff we want to pull out of the atmosphere. Carbon plays an important role in healthy soil. All life-forms contain carbon. All of 'em.

The carbon in soil comes in roughly three forms: living, dead, and really *really* dead.

- **Living** forms of carbon in SOM are live plants (mainly roots), microbes, and animals (bugs, worms, other crawly things).

- **Dead** carbon is basically compost in its various stages — dead plant and animal matter that is more or less recognizable depending on its stage of decay. This also includes little Fluffy, your kid's hamster that was buried in the backyard last summer.

- **Really *really* dead** carbon is what is affectionately known as humus, a stable form of carbon. This is the fine brown stuff, no longer recognizable as its original ingredients, that you see in compost.

As organic matter is digested and passed along the soil food web (we'll talk about this web in a moment), the carbon it contains is tacked onto an ever-growing chain — kind of a carbon bunny hop. When there are no more nutrients to digest, the end product is humus, the brown substance that gives good soil its characteristic dark color. Humus can be a stable carbon "sink" for eons — especially if it is not disturbed.

Tilling (this includes rototilling, turning with a shovel, or plowing) destroys soil life and causes soil to lose carbon faster than it accumulates it. It creates greenhouse gases while degrading soil quality. Turn your soil judiciously, if at all. I'll talk later about how to garden without tilling; this is what allows soil to get really rich. It will also save you a boatload of work.

WHAT'S LIVING IN MY BACKYARD?

Here's a look at the soil food web's cast of characters. A TEASPOON of healthy soil contains:

- More microbes than there are people in the United States
- More species than all the vertebrates on Earth
- Several yards of fungal filaments
- Several thousand protozoa
- Several dozen nematodes

In your quest for good soil, the most important thing you can do is *add organic matter*. The second most important thing is *don't till*.

So, now that you've got lots of studs (decomposing plant matter, not men, silly), how do you nail them together?

In your quest for good soil, the most important thing you can do is *add organic matter*.

The Soil Food Web: A Gardener's Best Friend

Maybe you've been hearing some buzz about the soil food web, nodding and smiling as though you understand it but feeling a little mystified. I'm about to help you with that.

If you're like me, you love gardening for the beauty and food it produces and for the sheer pleasure of being outside and watching stuff happen. This is a process we can do instinctively, often with great success. But my enjoyment of gardening increased dramatically once I learned more about the science of soil — the life in my dirt and on the plants growing in it.

Understanding the soil food web can inspire you to garden more sustainably. You feed the soil by using resources in the present without stealing them from the future. With this approach to gardening, you constantly improve your soil. And grow some bodacious vegetables.

The soil food web's community of organisms, pretty much all of which eat and are eaten by each other, comprises bacteria, fungi, archaea, nematodes, protozoa, arthropods (bugs), worms, and other small things, some of which we don't know about yet. Creatures living on top of the soil — beetles, spiders, and birds, for instance — also figure in. And *all* living things, from microbes through plants and right up to humans, contribute by eating, dying, and decomposing. You're welcome.

Together, the members of the soil food web keep the planet running by growing plants. Like little microscopic sponges, microbes and their soil food web friends make aggregates, or crumbs — clusters of soil particles that absorb water and create air spaces in soil. Soil aggregates, along with plant roots, hold water in the soil and also allow it to infiltrate and absorb into deeper ground, thereby preventing flooding and erosion. And they put the "crumb" in crumbly soil.

The soil food web cycles nutrients, eating or decomposing other living things and turning them into fertilizer plants can absorb. And it suppresses disease; all organisms "eat or compete" — that is, eat

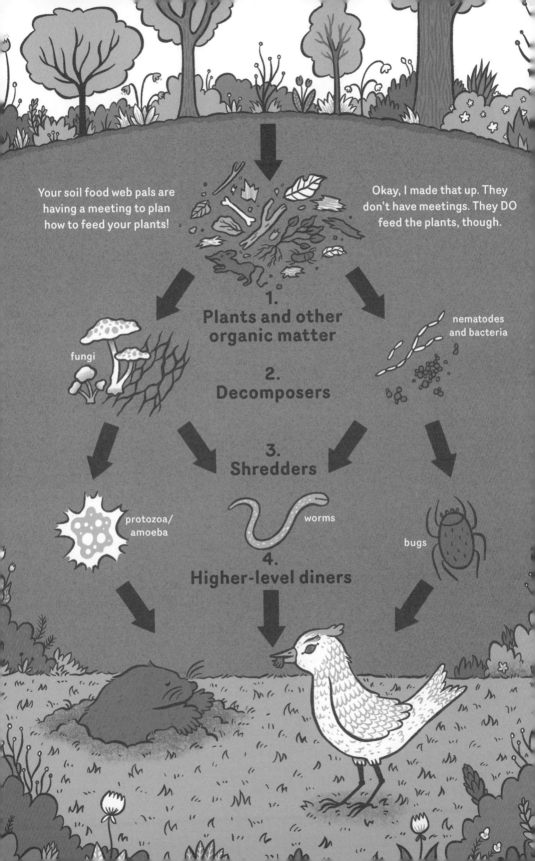

other organisms or compete with them for the food supply. A large and varied population of soil life keeps the bad guys from taking over.

This food web also pulls carbon out of the air, keeping it in the soil in the form of humus, which is the long chain of carbon molecules that makes rich soil dark. Humus is the end product of organic matter that has been serially digested by various soil residents; it holds water and functions as a sort of carbon condo for water, air, microbes, and nutrients, which are then available to plant roots.

Tilling disrupts the soil food web by chopping up worms and fungal strands, exposing organisms to air and heat, and letting soil carbon combine with oxygen, making carbon dioxide. And manufactured chemicals can damage inhabitants of the soil or kill them outright. In later chapters, I'll talk about how to protect and nourish your soil and the things living in it.

In short, the soil food web feeds everything you eat and helps keep your favorite planet from getting too hot. Be nice to it.

Biodiversity 101: Microbes

Soil microbes are the smallest inhabitants of your backyard — so small that you can't see them without a microscope. They have many functions, including competing with pathogenic (disease-causing) microbes, fixing nitrogen, making organic matter into humus, breaking down nutrients into usable form, and producing antibiotics to protect plants (and, sometimes, us). Have you ever used streptomycin, neomycin, erythromycin, or tetracycline? Thank a germ. A type of bacterium known as an actinomycete, to be precise.

The more I learn about soil microbiology, the amazeder I get. Microbes are a vital part of the soil food web, a mind-bogglingly complex system that is orchestrated by plants. Using photosynthesis, plant roots produce sugary exudates — think "sweat" — that attract the microbes a particular plant wants to work with: the ones that will adjust soil pH to the plant's liking and bring the right mix of nutrients. These microbes include bacteria, fungi, archaea, and who knows what else? Science is still learning. Always.

FUN FACTS ABOUT MICROBES

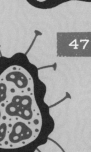

- Microbes and other soil organisms are capable of producing all the nutrients plants need. That's what they're there for.

- A teaspoon of good soil contains more microbes than there are people in the United States.

- The mass of bacteria on Earth is greater than that of all plants and animals combined.

- Actinomycetes, a type of bacterium that has some features of a fungus (spore formation, mainly), produce antibiotics that end in -*mycin*. They also are responsible for the wonderful smell of healthy soil.

- Ninety percent of cells contained in the human body are microbes(!).

- Microbes produce hormones that regulate growth, stress response, and immunity in plants and animals.

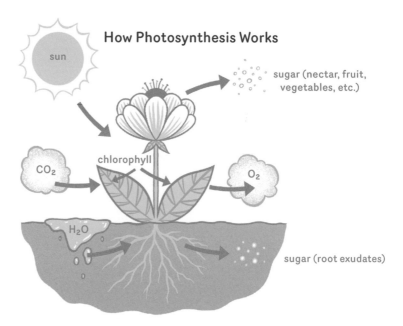

All these organisms predigest nutrients for plants; microbes, being sticky, hang around in the rhizosphere (root zone), mineralizing nutrients — turning them into a form plants can absorb. This is why the soil around roots is richer than the surrounding soil. Plants are in charge of a system that delivers nutrients to their roots, thereby enriching the soil they live in.

Microbes also form a shield around the roots, duking it out with organisms that could do the plant harm. Microbes function as sort of a sketchy plant bodyguard, shielding the plant, killing its enemies, and dishing out growth hormones, all while prechewing its food.

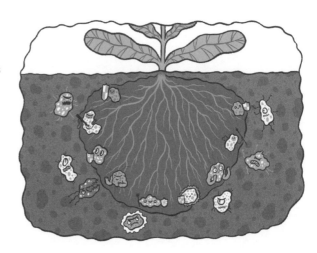

Friendly microbes are keeping out the riffraff. A microbial biofilm creates a sort of armor around plants (and us). They need us, so they protect us.

FUN FACTS ABOUT ROOTS

- Some grow deep into the ground, and some are "oblique" (angling down from the plant's crown). But much of the root growth is up, into topsoil or mulch, and out, to extend the radius of available nutrients. This is what makes mulching beyond the outer edge of plant foliage so important.

- Root hairs multiply root surface area exponentially. As they insinuate their way into even the hardest soils, roots blaze trails for other soil life. The channels they create make way for water, air, worms, nutrients, and the microbes that break down and anchor those nutrients.

- Roots "mine" the deeper soil for nutrients that wouldn't otherwise be available closer to the surface. These nutrients then remain near the surface for future inhabitants.

- When they die, roots add huge amounts of organic matter to the soil, continuing to feed the life they've attracted.

Why We Love Microbes

Fun fact about microbes: you love them, by and large. You especially love them if you're a gardener. Here's why:

1 Microbes decompose stuff. Decompose: to rot. Or, in fancier terms, "to break down organic matter, such as dead plant or animal tissue, into smaller organic particles that are available for use by the organisms of an ecosystem." In other words, microbes break down dead things to feed live things. Without microbes, there would be no soil, only rocks and sand (little pieces of rock).

Okay, granted, the whole "rotting" thing doesn't sound that lovable, but think of this: if it weren't for microbes, we'd be living on top of layers of perfectly preserved ancient redwoods, dinosaurs, and woolly mammoths. Plus cavemen and cavewomen, and more recently departed plants, animals, and ancestors, over whom we'd have to climb. I think you'll agree that's a little creepy.

2 If microbes weren't living in our digestive tracts, we could eat as much of those perfectly preserved meats and vegetables as we wanted, but they would be indigestible. They would pass through us perfectly preserved, just chewed up to varying degrees, and we would starve to death. Similarly, without microbes in soil, plants would be languishing in sand devoid of humus and nutrients.

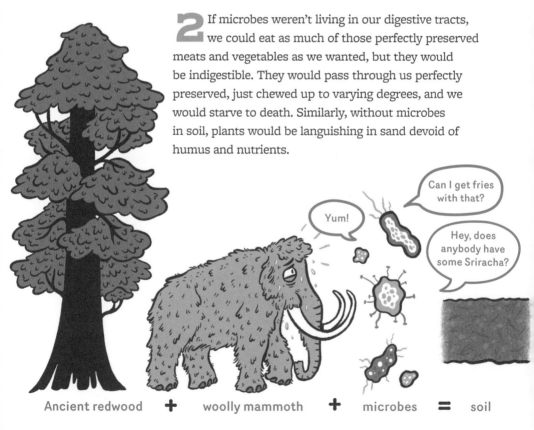

Ancient redwood + woolly mammoth + microbes = soil

Wait a minute — there's a flaw in my logic. If there were no microbes, life on Earth would never have existed, and there never would have *been* cavemen or woolly mammoths or trees.

3 (The real truth.) Without microbes, there would be no life. Earth would be a barren wasteland of a sphere hurtling through space, and you wouldn't exist. I know, right? Bummer.

Here's the take-home message: We need microbes, and the vast majority of them are beneficial. We especially need aerobic microbes — the ones that live where there is good aeration; they do much of the work in soils, as well as in and on plants and animals (that includes us; think probiotics).

So microbes break down dead things into food for live things. In our intestinal tract, a healthy mix of microbes (we're talkin' billions here) breaks down what we eat into absorbable form, produces nutrients, and keeps disease-causing microbes in check. On our skin, an army of microbes keeps the bad guys at bay. And in the garden, a diverse population of microbes in the soil, and in and on plants, will nourish and defend plants from diseases and pests. Our goal, always, should be to not eliminate anything but to add diversity.

So . . . we like microbes. I think everybody would agree, though, that we like some more than others. But what makes a good germ?

Well, it's all a little arbitrary. Since we're humans and we invented words, we get to use the word *pathogen* and define it as "something that causes disease" (*pathology gen*erating, get it?). I suspect a pathogen would think it's just doing its job, but if it causes diseases in us or other things we care about, we consider it a bad guy.

Trying to kill off bad guys is fruitless and destructive. What we want is a neighborhood full of all kinds of microbes, most of which will typically be nice ones. Imagine a town full of happy families with a few derelicts and criminals. Now imagine that town wiped out by chemicals; who do you think comes back first? Not the happy families. The war lords, who will loot JCPenney and Safeway and be dressing smart, eating whatever the heck they want, and having drunken parties and making babies. So it is with pathogens (disclaimer: this is my personal interpretation of the science of pathogens). Keep the bad guys surrounded by good guys, and the evildoers will be better behaved.

Most of the good guys in the garden are aerobic, meaning they need oxygen to survive. And most of the ones that cause us problems are anaerobic, meaning they don't need oxygen and often can't tolerate it. If we create a pleasant, airy, nurturing neighborhood, more nice microbes will come.

A good rule of thumb is that we want lots of, and many different kinds of, microbes around us. Let them duke it out — most of them are on our side.

Meet the Microbes

Three main types of microbes live in soil: bacteria, fungi, and archaea.

BACTERIA are the most abundant organisms in your soil and on your planet — they're small, even for microbes, and are fast reproducers under ideal circumstances. They've been around roughly a couple of billion years.

FUNGI are newer (measured in millions of years rather than billions) and more organized than bacteria. They group together into molds, mushrooms, and a giant network (more than 2,000 acres) of mycelia in a forest in Oregon — one huge organism that produces *Armillaria* mushrooms once a year. And kills trees, alas.

ARCHAEA are the oldest known organisms; they've been around for almost 4 billion years. Scientists used to think they were a type of bacteria, but no. Although archaea and bacteria look similar, they're genetically different. Scientists figured this out very recently, as science goes — in the 1970s.

Other soil microbes exist, such as rare hemimastigotes like the species *Hemimastix kukwesjijk* discovered by researchers in Nova Scotia in 2018. *H. kukwesjijk*'s role has yet to be understood by scientists, but its discovery goes to show how much there is still to learn about soil.

Let's take a closer look at the microbes we know.

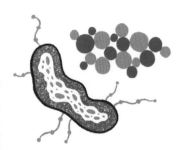

LESSONS FROM C. DIFF

Trying to eliminate a troublesome organism is far less effective than promoting a healthy variety of them. There is speculation that AIDS, Ebola, and other destructive viruses were kept in check by the incredibly diverse rain forests they lived in, until those rain forests were cut down. In diverse populations of microbes, bacteria and fungi eat each other and/or compete for the same nutrients ("eat or compete"), keeping any one species from growing out of control.

In human microbiology, consider *Clostridium difficile* (*C. diff* to friends) — a nasty and sometimes lethal intestinal bug. *C. diff* causes diarrhea in people who have received multiple courses of antibiotics, which kill off much of the normal intestinal flora. Traditionally, the treatment has been to throw more and stronger antibiotics at it. This has been met with limited success.

The most effective treatment is a fecal transplant, which is sort of a sourdough starter for the colon. *One teaspoon* of a healthy person's intestinal fluids contains roughly one billion microbes, in 20,000 to 30,000 varieties. Adding these back into a sterilized gut is the best way to control overgrowth of *C. difficile*. *C. diff* doesn't have to be destroyed — it just has to learn to get along with the neighbors.

REPRODUCTION, TWO WAYS

Asexual reproduction is more efficient than sexual, because each cell doubles itself with each cycle of reproduction; sexual reproduction requires two cells to make a single offspring. Remember that old question "Would you rather have a penny that doubled every day or 10 million dollars?" Correct answer: the penny, which would turn into $10,737,418.24 after 30 days. This assumes you want to be as rich as possible. Which, in my opinion, is overrated. But I digress...

The fact that sexual reproduction is the hallmark of higher organisms is something of a mystery; why haven't cells that constantly double in number — the more primitive ones — taken over the world? The theory is that because sexual reproduction combines the genes of organisms that are best at surviving, the offspring has the advantage in "fitness" to live.

BACTERIA

Bacteria are the second oldest form of life, after archaea. A few billion years ago, a type of bacterium called cyanobacteria began to photosynthesize, pulling carbon dioxide out of the air and turning it into carbohydrates and oxygen. After another billion years or so — a period dubbed "the boring billion" by scientists — oxygen concentrations became high enough to support animals, and have now leveled off at about 21 percent of the atmosphere. We're indebted to cyanobacteria (which are often mistakenly called blue-green algae but are actually a type of bacterium); plants still work symbiotically with them to photosynthesize. Enjoying that oxygen you're breathing? Thank cyanobacteria.

Since bacteria reproduce by cell division, their population can grow at a dizzying rate (remember that penny example?). A single bacterium could potentially produce about five billion offspring in 12 hours.

Not surprisingly, then, bacteria are the most numerous residents of the soil food web. According to the U.S. Department of Agriculture, a teaspoon of healthy soil contains somewhere between one million and one billion bacteria. That's as much mass as two cows per acre. And those bacteria comprise tens of thousands of species, all working together to feed the soil.

What are all those bacteria doing? Here are some ways they serve the soil food web:

DECOMPOSERS. Bacteria, along with fungi, are the primary decomposers in soil, breaking down organic matter into nutrients for plants. After they dine on the nitrogen, carbon, and other nutrients from that organic matter, they hold on to those nutrients until they die, generally not of old age but when they're eaten by other organisms. In the process, nitrogen and other nutrients are released into the soil, where plants gobble them up. Bacteria are an important food source for other members of the soil food web.

Bacteria prefer fresh, green organic matter. Fungi, in contrast, like brown, dead stuff. Bacteria are the reason your compost pile heats up when you add grass clippings; the ensuing bacterial feeding frenzy produces heat.

NITROGEN CYCLERS. Nitrogen is constantly moving around, from the air to the soil to serving as building materials for living things. Bacteria play an important role in this nitrogen cycle.

Some bacteria obtain nitrogen by decomposing organic matter; others pull nitrogen out of the air. Bacteria also convert soil nitrogen into different compounds, some of which plants can use. And still others convert nitrogen back into the form it takes in the atmosphere. Since nitrogen is perhaps the most important plant nutrient, a diverse bacterial population is vital to healthy soil. Bacteria feed other soil organisms, which feed plants either with their wastes or by dying and decomposing.

AGGREGATE FORMERS. Bacteria produce sticky stuff called bacterial slime, which helps them stay in their chosen neighborhood. This slime sticks smaller soil particles together into aggregates, or crumbs. Those crumbs allow water and air to pass between them, and they also store some of that water where it's available to plant roots.

NUTRIENT IMMOBILIZERS. Nutrients in soil are either immobilized, when they're contained by a living organism, or mineralized, when they are released as waste or by decomposition into a plant-available form. Bacteria (and other soil organisms) hold nutrients in place, preventing them from being leached out of soil.

Because bacteria are happiest in the root zone, where they are fed sugary exudates, the soil in that area grows increasingly rich and crumbly.

DISEASE SUPPRESSERS. All soil organisms eat or compete with other organisms for food. Thus, a healthy mix of life in soil keeps any one pathogen from growing out of control. Bacteria both directly compete with pathogens and feed a diverse group of other soil organisms; these processes maintain a healthy balance of soil life, keeping any one thing from dominating.

Some bacteria also produce antibiotics. For example, actinomycetes, the bacteria that give good soil that wonderful smell, produce streptomycin and other antibiotics commonly used to treat humans. Antibiotics produced by bacteria help control pathogens both in humans and in soil, where they keep diseases in check.

FUNGI: THE SHORT COURSE

Fungi were one of the first eukaryotes — more complex cells that have their DNA enclosed in a nucleus. (Prokaryotes, in contrast, are usually single-celled organisms whose DNA is just floating around in the cellular fluid.)

Probably the most famous fungi are mushrooms, mildew, athlete's foot (tinea pedis), and yeast infections. One out of four has a fan base; the rest, not so much. Fungi, however, are a huge part of the soil food web. In addition to being primary decomposers, they feed plants and improve soil structure.

The most basic form of fungus is a one-cell-thick strand called a hypha (plural: hyphae), which sprouts from a spore, basically a mushroom seed. Individual spores and hyphae are invisible without a microscope. Those white filaments growing on rotting logs (mycelia), and that fuzz you find inside takeout boxes in the back of your refrigerator, are composed of hundreds of thousands of hyphal strands growing alongside each other. This bundling helps them conserve and transport moisture within the strands.

FUN FACT ABOUT MUSHROOMS

Mushrooms are the "fruiting body" of mycelia. They exist to spread spores to new territory, in the same way flowers and fruits exist to make seeds. Thus, picking a mushroom isn't like cutting down a tree — it's more like picking an apple.

Although you can't see them, a shovelful of healthy soil holds *miles* of hyphae, busily forming mycelial strands, improving soil texture, and breaking down nutrients so plants can use them. In the book *Teaming with Microbes*, Jeff Lowenfels and Wayne Lewis write that fungi can expand up to 40 micrometers a *minute*, as compared to bacteria, which may travel only 6 micrometers in a lifetime. You may have observed this speedy fungal growth when you see mushrooms pop up, literally overnight.

Fungal strands act as the soil's circulatory system. Because they can move stuff through their cell walls, they transport moisture and nutrients from a food source they're breaking down — rock, dead logs, rotting plants, or insects — to a location yards away, where other organisms can use them.

Some mycelia, called mycorrhizae (literally meaning "fungus roots"), intertwine with plant roots in a mutually beneficial relationship. In some forests, a single mycelial network can extend for acres — one huge underground organism supporting thousands of trees.

So, fungal hyphae (those single-cell-thick strands) form mycelia (bundles of those strands), and, under the right conditions, grow into fruiting bodies such as puffballs, truffles, and mushrooms, among others. Mushrooms exist to disperse spores, like an apple exists to distribute seeds. And sometimes to be delicious.

MYCORRHIZAE. These fragile and vitally important fungi are the main reason you should sell your rototiller. Mycorrhizae are a type of mycelium that intertwine in or around plant roots — think hair extensions. This is a win-win situation: These root extensions enlarge the range of plant roots many times, bringing moisture and nutrients from farther than the root itself would ever grow. And the mycorrhizae, in exchange, use sugars the roots excrete (plants are in charge, remember?) to nourish the rest of the fungal organism. Almost all plants work with mycorrhizae.

Some plants and mycorrhizae are a little . . . how can I put this delicately? . . . promiscuous, while others are specific to each other and will only work with a certain type. Case in point — the Puerto Rican pines (see opposite).

The story of the Puerto Rican pines astounds me. Its implication is that the mycorrhizae had evolved in South Carolina, probably over eons, to work specifically with pine trees. These mycorrhizae didn't exist in Puerto Rico's rich and diverse ecosystem, and apparently the mycorrhizae present couldn't adapt over 2 decades of languishing pine production. Nutrients in the soil were unavailable to the pine trees unless predigested and delivered by the mycorrhiza specific to that tree.

More than 50 years later, the mycorrhizal pines are still growing strong and naturally reproducing in the pine plantations. Most of the plants you grow are equally dependent on mycorrhizae.

THE PUERTO RICAN PINES: A LESSON IN THE IMPORTANCE OF MYCORRHIZAE

In the 1930s, various attempts were made to plant pine forests in Puerto Rico. This was a type of meddling common in those days — the U.S. Department of Agriculture Forest Service brought in non-native species because they thought Puerto Rico needed them. The "Pines to Puerto Rico" adventure provided the beginning of our understanding of mycorrhizae, which evidently evolve with the plant that depends on them.

For almost 2 decades, despite much tinkering by the Forest Service, the pine seeds brought to Puerto Rico would germinate well and grow to a height of about 4 inches, then wither and die. The Forest Service tried adding phosphorus to the soil, thinking the trees couldn't develop due to a deficiency. No dice; the trees still died in infancy. No manner of fertilizing and nurturing seemed to help.

Then, in 1955, B. J. Huckenpahler of the Forest Service brought in native soil from South Carolina, the origin of these pine trees. Rightly thinking that the trees were missing something ineffable from home, he mixed the Carolinian soil into the dirt around 32 seedlings and left a second set of 32 alone as a control.

A year after the addition of native soil, the inoculated trees were about 5 feet tall. The other 32 were mostly dead. Further research showed that mycorrhizae were the key ingredient in the South Carolina soil. The formerly pining pines (sorry . . . I had to do it) flourished.

So now that you're a mycorrhizae aficionado, how do you get some for your garden? You can purchase an all-purpose mix at some nurseries. This mix will work for many garden plants, with the exception of the cabbage family and "Ericas" (rhododendrons, blueberries, laurels), which, for some reason, seem to have no use for mycorrhizae. You'll probably see the words *endomycorrhizae* and *ectomycorrhizae* on the label; endomycorrhizae actually grow into plant roots, while ectomycorrhizae wrap around them. A mix will most likely include something that works with your plants.

When using a mycorrhizae mix, roll moistened roots and even seeds in the stuff to give your plants a head start; the direct contact won't hurt, and it may well help. You can also stir some into the planting hole, or add it to compost. Even if the mix is not the fungus of choice for what you're planting, it will inoculate your soil with more life and benefit plants that arrive later.

The best approach, of course, is to create soil so full of life that the mycorrhizae are present already. This, however, usually requires the investment of some years, as you build up your microbial population by adding mulch and compost, growing diverse plants, and avoiding the rototiller. But I like to give my plants the benefit of the doubt with a little store-bought mycorrhizae.

ARCHAEA

In the late 1970s, scientists were surprised to discover a whole new branch on the tree of life, distinct from bacteria and eukaryotes: archaea. Although they look like bacteria, there are distinct differences, including what their cell walls are made of, their sources of food, and their genetic makeup. Archaea are more closely related to eukaryotes, those more complex cells that have their DNA enclosed in a nucleus.

Archaea were first discovered in extreme environments, such as hot springs and geysers, and named "extremophiles," but we now know they're everywhere, including in soil as well as in our digestive systems. But because of their ability to survive heat, cold, salt, and acids — harsh environments similar to early Earth — they're believed to be the oldest organisms on the planet.

Archaea serve important functions in soil: They're involved in nitrogen and carbon cycling, helping make those nutrients available to plants. And they live much deeper in soil than bacteria, feeding deep plant roots.

Archaea also work as decomposers and produce a class of antibiotics distinct from those produced by bacteria, sometimes allowing them to work where other antibiotics are ineffective.

MICROBES AS SOIL REMEDIATORS

Here's yet another reason to love microbes: they remediate toxic soil. In a dramatic example of this, mushroom guru Paul Stamets (documented in his book *Mycelium Running*) inoculated a pile of petroleum oil–saturated soil with oyster mushroom spawn. Two similar piles had bacteria or chemical fertilizers added.

After 8 weeks, the pile inoculated with oyster mushrooms was covered with robustly growing mushrooms with no trace of hydrocarbons (in other words, they were safe to eat). And it was teeming with microbial and plant life; the other two piles looked like, well, slag heaps. Nature is a master of remediation. If we ensure that a wide variety of life is available, nature will take it from there. Nature's amazing like that.

The mushrooms also paved the way for other forms of life: the rotting mushrooms attracted flies and other insects, which attracted birds and the seeds spread by the birds.

Biodiversity 102: Algae, Bugs, Worms, and More

Microbes are just part of the soil food web — the part you can't see, at least individually. There's also a host of visible members that live in and on the soil.

Algae

Algae cells can be either prokaryotic (primitive, with no cell nucleus) or eukaryotic (containing a nucleus — more highly evolved). As such, algae could be considered a transition organism, bridging the gap between early cells and later ones. Algae are, in fact, the earliest form of eukaryotes, which require a higher level of oxygen than that on early Earth. Eukaryotes paved the way for higher forms of life, including fungi, plants, and animals.

The algae spectrum runs from organisms like the tiny one-celled *Chlamydomonas* genus, which can photosynthesize and is considered an ancestor to green plants, to giant kelp.

What does this have to do with your garden? Naturally occurring algae in soil can fix nitrogen — that is, pull nitrogen from the air and stick it into the soil where plants can use it. And kelp, the largest form of algae, is often used as a plant nutrient, either as a soil additive or foliar spray. It contains high amounts of potassium, as well as trace minerals, vitamins, and plant hormones. Kelp has been shown to increase root mass and help with resistance to disease.

Another form of algae, diatoms, have shells made of a glasslike substance; their remains are mined and sold as diatomaceous earth, which is used to control garden pests like slugs, snails, and problematic insects. Diatomaceous earth helped me control my population of harlequin bugs, which were decimating my broccoli and kale.

Protozoa

Protozoa (amoebae and others) are the first animals we'll meet in the soil food web, the smallest animals (one cell), and the first animals to evolve; their name comes from *proto* (first) and *zoon* (animal). They are single celled but much larger than the bacteria and fungi on which they dine — so much larger that they're sometimes visible with the naked eye. After digesting microbes and other protozoa, they release nutrients from the consumed organisms as wastes in a plant-available form. Interestingly, since protozoa contain less nitrogen than do the bacteria they eat, they release that extra nitrogen into the soil, conveniently located in the root zone where they like to live.

Protozoa participate in a system of checks and balances with bacteria. When bacteria are plentiful, protozoa dine well and multiply. As they reduce the population of bacteria and therefore their food source, protozoa decrease in number; bacteria then multiply more freely, protozoa increase, and so on; lather, rinse, repeat.

Nematodes

Nematodes are tiny roundworms, usually visible only with a microscope. Since they can live in or out of water and in a wide variety of environments, they are one of the most abundant forms of animal life on Earth, right up there with arthropods.

According to Nathan Cobb, the father of nematology, in his book *Nematodes and Their Relationships*:

> If all the matter in the universe except the nematodes were swept away, our world would still be dimly recognizable, and if, as disembodied spirits, we could then investigate it, we should find its mountains, hills, vales, rivers, lakes, and oceans represented by a film of nematodes. The location of towns would be decipherable, since for every massing of human beings, there would be a corresponding massing of certain nematodes.
>
> Trees would still stand in ghostly rows representing our streets and highways. The location of the various plants and animals would still be decipherable, and, had we sufficient knowledge, in many cases even their species could be determined by an examination of their erstwhile nematode parasites.

Crazy, right?

Different nematodes are designed to eat bacteria, fungi, plants, and other animals like protozoa, grubs, and, lucky for the gardener, slugs. Predator nematodes also eat each other, so a healthy population of them keeps destructive nematodes in check. Some nematodes are harmful to plants or animals (the particularly nasty Guinea worm that lives in tropical climes is an example), but most of them are benign or even beneficial. For example, corn plants being attacked by rootworms send out a distress signal that nematodes sense; the nematodes then head over to the corn root and infect the rootworms. And one type of nematode is even sold as a biological control for slugs and snails.

The main reason we love nematodes is that they mineralize nutrients; this means they eat other organisms and release the nutrients in their wastes. Since nematodes need less nitrogen than protozoa, more of it is released into the soil, to the benefit of plants.

Nematodes also help the soil food web expand. As they travel around at relatively breakneck speeds, they carry bacteria and fungi on their skin to other parts of the soil, thus expanding the benefits of those organisms.

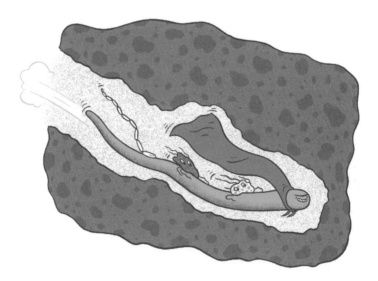

A nematode is racing around your soil while mineralizing nutrients. Go, nematode!

Arthropods

Arthropods, otherwise known as bugs, are outnumbered by nematodes but have a greater variety of species — about 1 million, as opposed to 20,000 to 40,000 known species of nematodes. They're characterized by an exoskeleton (hard shell), segmented body, and paired, jointed legs.

The following are types of arthropods: arachnids (spiders, scorpions, mites, ticks, daddy longlegs), centipedes, millipedes, insects, sow bugs and pill bugs, and some sea creatures. They can be microscopic or large and delicious, as in the case of king crab and lobster.

Arthropods play an important part in the soil food web, living on or in soil. They shred organic matter, passing it on to smaller organisms and thereby increasing the populations of fungi and bacteria. They eat those fungi and bacteria, releasing the nutrients contained in their bodies into the soil. And arthropods transport microbes that hitchhike on their bodies, thus expanding the microbes' range farther when those arthropods are eaten by bigger predators.

While going about their daily business, arthropods help break down organic matter and aerate the soil. Although some are destructive, most are beneficial to soil. As with all members of the soil food web, a large number and variety of arthropods is a good thing.

Worms

Worms are perhaps the best-loved members of the soil food web. There are three rough categories of earthworms:

o **Epigeic ("upon the earth"):** These worms live on the surface as opposed to burrowing down into the soil, and are the ones most familiar to gardeners who compost.

o **Endogeic ("in the earth"):** These worms live near the surface and are great soil tillers. They tend to be pale in color — gray or light pink — and not striped. They travel to and fro just below the soil surface, digesting and mixing in organic matter.

WORM ANATOMY: BECAUSE IT'S AMAZING

Earthworms can be thought of as a tube within a tube, since they have one long, straight digestive tract that runs head to toe. But they are *waaay* more complex than a simple digestive tube. They have a nervous system that smells food and detects light and salt, both of which they dislike (which is why they writhe when exposed to sunlight or skin). They have a circulatory system with several little hearts. And they have bristles that anchor their more forward parts while allowing their nether parts to catch up — this is how they move through soil. In place of a skeleton, worms use a fluid-filled chamber that surrounds their digestive tract; they're sort of like long, skinny water balloons that fertilize soil and crawl. And those are just their highlights.

All earthworms are hermaphrodites, meaning they have both male and female organs. Their lives are not completely devoid of romance, however — they need to mate with a partner to make babies. They do this by lying alongside each other, head to tail, and joining their respective male and female parts. They then each form a sort of shell over that chubby midriff you see on mature worms, called a clitellum (this is where the female parts and eggs reside). Inside that shell, eggs and sperm are mixed together. The worm then sheds the shell and fertilized eggs, leaving a sort of onion-shaped cocoon. You'll notice these in your compost, now that you know what they are.

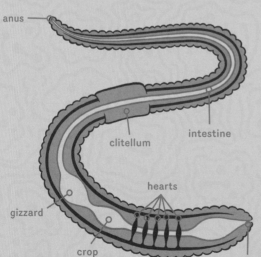

- **Anecic ("out of the earth"):** These worms have a tendency to appear on the surface, although they make their homes deep in the soil. Anecic worms are also known as nightcrawlers because they surface at night, for a couple of reasons: to get organic matter to eat, and to mate. They do these things at night because the surface is cooler and moister then, and therefore less likely to dry them up.

Eisenia fetida, also called red wigglers, are epigeic. They're the ones you find in your compost pile and wiggling on the surface when you move loose organic matter. Being prolific breeders, they're invaluable in vermicomposting. They are, not surprisingly, red: if you look closely, you'll see that they are also striped.

The common earthworm most of us are familiar with is *Lumbricus terrestris*, Latin for "earthworm of the earth." They are anecic. Also known as nightcrawlers, these worms create deep vertical burrows, generally one per worm. These burrows create channels for water, air, and roots to enter the soil. When surfacing for food, nightcrawlers mix in organic matter from the surface and turn it into castings.

The common earthworm was introduced to North America by European settlers. Although beloved by gardeners, it can be detrimental to forests; because of its voracious appetite for organic matter, it changes the ecology of forest floors, making life tough for new growth. This is why you shouldn't dump your nightcrawler bait in the woods when you're done fishing.

All worms are harmed by tilling and chemical fertilizers. That's why I garden the way I do. I have a love for worms that borders on unnatural.

Slugs and Snails

Although they are one of God's creatures, I waste no love on slugs. Nor snails. The lack of affection isn't mutual, alas — these mollusks love my moist, mulched garden.

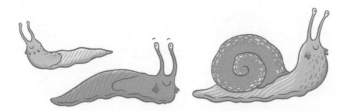

Snails and slugs do have a function in the universe besides demolishing new seedlings and breaking my heart. They also shred organic matter, thereby helping decompose it. And they provide food for lizards, toads, and other things in the garden that I actually *do* love. Mostly, however, I want to kill them, which I do by scattering iron phosphate (an organic snail bait) or by laying traps in the garden under which they gather and have little slug meetings to plot how to ruin my life. These traps can be boards, grapefruit rinds, or thick leaves like those from cabbage. In the morning, I turn over the traps and stomp on the mollusks. I apologize for this graphic violence in an otherwise pacifist book. I'm happy to coexist with slugs and snails outside my garden; I *don't* welcome them in to dine on my plants.

Other steps that help control mollusks are spreading diatomaceous earth, the sharp edges of which injure snails' soft underbellies; encircling raised beds with some form of copper (tape, wire, pipe), which gives snails an electrical shock; and watering only early in the morning, so the garden dries out during the day and is moist for fewer hours. Also, in areas where they're particularly troublesome, you can use wood chips for mulch; these tend to be drier and rougher, therefore slug-unfriendly.

The use of coffee grounds to control slugs is a myth — I have an action-packed video to prove it. Beer traps reputedly work; I haven't had the required diligence, but you might.

Things That Live on Top of the Earth

Creatures that live aboveground, like birds, lizards, snakes, toads, mice, gophers, and rabbits, also figure into the soil food web. Though they may or may not be welcome in your garden, they fill an important niche. All of them produce manure that feeds plants. All of them indicate that you have soil food web activity (something for them to eat). And all of them fit into an ecosystem that we'll never fully comprehend. If they're causing you grief, try to use mechanical controls (traps or barriers). Often, though, these critters can be beneficial. Snakes can eat slugs, gophers, and other pests. And western fence lizards helped me vanquish the harlequin bugs that were decimating my kale and broccoli (see The Harlequin Bug Caper, opposite).

THE HARLEQUIN BUG CAPER

A couple of years after I moved into my house, harlequin bugs discovered my cruciferous crops. They were all over my kale, broccoli, and mustard plants, leaving them covered with hard, inedible spots as though the plants had measles.

An application of diatomaceous earth helped quite a bit, cutting the population by maybe half. The next year — not coincidentally, I think — there was an explosion of fence lizards, which eat beetles. I would wander the garden in the evening when the bugs were shamelessly copulating under the setting sun and knock them off the plants. The lizards took care of things from there. I've only rarely seen harlequin bugs since. This is how I like it — just a few pests in the garden to feed the guys that eat them. And this is how nature works, if you let it.

DO NO HARM

The best thing we can do for the soil food web, and for the planet, is to garden gently. Besides not tilling, it's also important to avoid chemicals. A healthy soil food web will control pests and provide all the food your plants need. Chemical fertilizers will damage that soil food web by killing indiscriminately.

I use only organic pesticides or fungicides, when I use anything, because I love bees and nature and my grandkids. And the human race, by and large. If anything can bring about real change, it's our love of children. Let's consider who'll be living in the world we're creating. Let's do better.

CHAPTER 4
INSTALL VENTILATION, PLUMBING & A NICE PANTRY

No-Till Growing, Paths & CEC

Your soil house will need a steady supply of fresh air and water, and a place to store nutrients. The best way to provide these is by nurturing and protecting the soil food web. Biologically active soil turns soil particles into larger crumbs; these allow water to be absorbed and the excess to leave by means of gravity. The resulting vacuum will pull air into the newly empty spaces. Voilà! Plumbing and ventilation!

Your soil's cation exchange capacity (CEC) tells you how well your soil can store nutrients. CEC is basically a system that uses electrical charges to store and release nutrients (more about this shortly); it's your soil's pantry. If your soil has a low CEC, any nutrients you add to the soil will wash out after giving plants a quick binge. The best way to raise CEC is to add compost and biochar (charcoal) and increase biological activity.

Step Away from the Rototiller!

I was once like many of you. I itched to till those nice grasses and weeds into my garden soil, to fluff it up and nourish the coming tomatoes. Then I went over to the "dark side" — dark as in rich, dark, carbon-filled soil.

It's true that rototilling gives soil microbes a springtime binge on green matter, and it's sometimes helpful for already dead soil. It also, however, kills or removes lots of things that you want to live in your soil. To wit, your innocent-looking rototiller is disturbing the following:

MYCORRHIZAE. I think of these long fungal strands that we met in chapter 3 as soil's circulatory system. They twine around plant roots and extend their reach farther and faster than roots could grow — sort of like living hair extensions. They then bring back nutrients and water the roots wouldn't be able to reach otherwise, sometimes from as far away as 250 yards!

Like some panhandlers, mycorrhizae will work for food; they exchange nutrients and water for carbon sugars that plant roots exude. In the process, they further digest the carbon into more stable, darker hunks of humus.

WORMS. You may love or hate worms, depending whether you're:

o A 10-year-old boy making a squeamish girl scream

o That screaming girl

o A compost goddess (like me) with a love for worms that borders on unnatural

Worms eat decomposing plant matter, turning it into rich castings (known less elegantly as worm poop) and further stabilizing the carbon that plants have pulled from the air into the soil.

LIFE. Life *in* the soil creates life *on* the soil. That life — microbes, mycorrhizae, worms, bugs, and other very small things we have yet to discover — holds water and nutrients in soil, in a form plants can use.

Tilling chops up soil life and lets water evaporate, making soil dry and lifeless; think of tilled fields you've seen. Gray, dry dirt, am I right?

These two worms are about to reproduce in undisturbed soil.

Healthy soil needs water, organic matter, and to be pretty much left alone, so the life we want can flourish there.

Life puts the "crumb" in "crumbly soil." We want it there because, as we've learned, it's part of a long, *long* food chain that sequesters carbon in soil and makes it healthy. This soil food web breaks down nutrients into smaller chunks that plant roots can absorb. In the process, it makes your clay or sandy soil into the moist, crumbly black stuff gardeners love: tiny clusters of carbon and other organic matter stuck together with bits of sand and clay by microbial slime, the same stuff that is absolutely necessary for healthy soil.

The more dry, exposed, and "fluffed up" your soil is, the less hospitable it is to life — including worms, germs, and your arugula.

CARBON. As in carbon dioxide, the greenhouse gas we're trying to make less of so we have a habitable planet to give to our grandchildren. Carbon is what humus is made of and what makes rich topsoil dark. Carbon enters the soil via the soil food web; as living organisms are serially digested, the digesters tack carbon onto ever-lengthening chains that eventually become humus. But carbon *leaves* the soil when your rototiller mixes carbon with oxygen and they marry, make carbon dioxide, and float up to destroy the world as we know it.

Natural Aerators

Scientists (and gardeners!) are only beginning to understand the world of soil biology. What we've learned is this: everything we've been doing is wrong. (It's a lot like child rearing, that way.)

Okay, not everything. Many well-intentioned organic gardeners, though, have been rototilling once or twice a year, mistakenly believing that chopping and mixing in weeds and other organic matter will fluff, aerate, and improve the soil. We now know that tilling damages the soil instead of improving it, in part by breaking up soil particles and turning them into dust.

Imagine pouring water through a colander full of Cheerios; the water wets the Cheerios and drains right through. Now imagine first grinding Cheerios to the consistency of flour, putting the flour in the colander, and pouring water over it. The result is glop, right? That's what a rototiller does to soil.

The transformation of soil crumbs into dust is, in large part, what caused the Dust Bowl. With the advent of mechanized farming in the 1930s, farmers tilled in prairie grasses that formerly anchored the soil and kept it alive. A combination of drought and wind then blew that pulverized soil from the Great Plains, sometimes as far as New York City.

Tilling in organic matter is a good thing for bad soil, but think of this as a rare or one-time event. If the soil you're starting with is compacted, graded, or generally lacking biological activity, tilling can start it on the road to health. Once soil is loosened, though, we need to foster and conserve soil life and let nature go to work.

Living soil is well-aerated soil. Air is brought into soil in many ways, by many life-forms. Earthworms dig holes as they eat organic matter. Mycorrhizae, protozoa, and arthropods aerate soil with their comings and goings. And bacteria knit small soil particles into larger crumbs that make spaces for air.

Our goal is to create a happy world for the soil food web by feeding it (mulch, compost) and keeping it moist and aerated. Air and moisture attract soil life, and soil life holds moisture and nutrients and aerates soil. It's a wonderful feedback loop.

POP QUIZ: HOW DO YOU AERATE YOUR SOIL?

ANSWER 1: Rototill it!
Bzzzt. WRONG. Usually. Did you even *read* the 10 ~~commandm~~ suggestions at the beginning of the book?
ANSWER 2: Encourage diverse biological activity!
Ding ding! We have a winner!

Once your soil is happy, pulling up spent plants and weeds is all the soil loosening you'll need to do. Those plants leave behind some root filaments (organic matter) in the soil. The place where the roots used to be is now filled with . . . wait for it . . . *air*! And when you mulch with the plants you've pulled up, they form a nourishing home for more worms and germs, which are the best things for your plants. I LOVE this.

Bottom line: the more stuff there is living in your soil, the better aerated it will be, and the harder it will be for any pathogen to take hold.

Make Paths

Walking on soil compacts it and results in hardpan, a nearly impregnable layer of hard dirt below the soil that you've turned and planted. Hardpan impedes both drainage and root growth. Acidic and heavily clay soils are more prone to forming hardpan, but it's largely created by us humans walking or driving equipment over areas we intend to cultivate.

Protect your soil from being trampled by installing paths that give you access to permanent planting beds. These beds should be small enough that you can reach any part of them from a path, without walking on the growing area.

You can make paths out of almost anything: straw, upside-down strips of carpet, gravel, or wood chips (my personal favorite).

If you're starting with a rectangular vegetable patch, this process is pretty simple. First, design the vegetable beds to be no more than 4 feet across (2 feet if they're against a fence) so you can reach individual plants without walking on their roots. Then plan the paths between each bed. A path 2 feet wide is good enough, 3 feet is luxurious and easier to navigate with a wheelbarrowful of mulch. You can delineate the beds by means of a structure (such as a raised-bed frame), a row of rocks or cut tree branches, or simply the paths themselves. In a larger, more free-form area, space paths no more than 4 feet apart, or put in stepping-stones that allow you to reach plants without stepping on cultivated soil.

You can make paths out of almost anything: straw, upside-down strips of carpet, gravel, or wood chips (my personal favorite) — whatever you can find and afford. Remember, though, that weeds invariably sprout in paths, so it's helpful if your choice of path surface allows for easy weed removal. Gravel has a nice, formal look, but it's more difficult to pull weeds from a gravel pathway. Wood chips make beautiful paths and can often be sourced for free from tree-trimming services or community brush recycling centers. Wood chips are also permeable, and as they decompose and are "refreshed" each year, they encourage soil life even as they form an attractive walkway. When it comes time to weed a bark path, simply use a hoe or hand cultivator to slice wayward plants off just below the path surface.

An old Japanese proverb says: "The best fertilizer is the footsteps of the farmer." Yes, the best thing you can do for your garden is to putter in it — unless you're trampling the dickens out of your soil. Make paths. Use them. Seriously.

What Is CEC?

CEC stands for cation exchange capacity, a measure of how many positively charged ions (nutrients) can be retained on negatively charged soil particle surfaces. In effect, CEC measures the soil's ability to hold on to and distribute nutrients in forms plants can use, by means of positive or negative charges. It's your soil house's pantry.

Here's how it works.

The minerals that plants use are all ions, either cations or anions.

SOIL CATIONS (positively charged ions) include calcium, magnesium, potassium, ammonium (the plant-usable source of nitrogen), hydrogen, and sodium.

SOIL ANIONS (negatively charged ions) include chlorine, nitrate, sulfate, and phosphate.

Soils that are sandy and/or low in organic matter have a low CEC, meaning they can't hold on to cations like nitrogen and potassium. You can't improve CEC simply by pouring on fertilizers; you need to add organic matter to give those nutrients something to hold on to. Raising the CEC will allow nutrients to stay in the soil instead of leaching out.

In contrast, clay soils have high CEC, but the nutrients may be packed in so tightly that they're relatively inaccessible to plants. Adding organic matter to clay breaks it into aggregates or crumbs, and allows anions to catch and hold on to cations.

As you see, whether you have sandy soil or clay soil, the most effective way to increase CEC is to add organic matter. We'll learn how in a moment.

Does this all sound too scientific? Fortunately, you don't need to understand CEC calculations to be a good gardener. As Anna Hess says in her excellent book *Homegrown Humus,* think of the cation exchange capacity as tables in a restaurant. An eatery with lots of tables (high CEC) and lots of customers (available nutrients) will hum along. A restaurant with one sole table (low CEC) will languish, no matter how many customers mill around, because the customers will move on to someplace with available seating. Things look equally dire for a restaurant with plenty of tables but no customers.

A healthy soil food web is like Nevada City, with lots of tables (high CEC) and a steady stream of tourists/customers (available nutrients) year-round. In contrast, a low CEC soil subjected to chemical fertilizers is more like 1969 Woodstock, when a flood of humanity overruns businesses and leaves the place trampled, depleted, and forlorn.

How to Increase CEC

Organic matter, clay, and biochar all make soil more nutritious while raising its CEC. They do this by grabbing, storing, and releasing nutrients electrically. Your goal should be to have a CEC value of greater than 20, and biologically active soil.

ORGANIC MATTER increases both the soil's pantry space and food to put in it. Organic matter stores nutrients in the form of humus and microbial bodies — tiny, living lunch boxes full of plant snacks. These microbes hang around in the root zone and do many wonderful things, including releasing nutrients where the roots want them. Have I mentioned how great organic matter is for your soil?

CLAY retains lots of nutrients but keeps them inside something resembling cement, until the soil food web breaks up the clay and mixes it into soil aggregates (crumbs).

BIOCHAR (see opposite) raises CEC because it has a lot of surface area. Biochar is a wonderful storage system, but it starts out pretty much sterile. You have to put some food into it.

You'll notice that I don't list chemical fertilizers as a means of improving soil fertility. Chemical fertilizers damage the soil food web, either by killing organisms outright, by irritating them so they leave, or by making them less necessary and eliminating the processes that create and feed them. Adding nutrients alone won't improve your soil, although it may let plants gorge themselves briefly. Instead, having lots of life in your soil ensures a steady supply of nutrients and provides the ability to store them (high CEC).

In chapter 6, we'll talk more about how to add organic matter, nutrients, and microbes with compost. But right now, let's dive into biochar — figuratively speaking — which is chock-full of CEC.

Biochar

Biochar is a stable form of charcoal formed by burning plant material while giving limited access to oxygen. If the process is done in a specialized pyrolysis device (one that burns in a high-temperature, oxygen-depleted atmosphere), it also produces fuel in the form of oil and gas. This method of burning turns wood into charcoal, as opposed to ash. Rich in carbon, biochar is considered one route to carbon sequestration (storage).

The power of biochar was first discovered by early European settlers in Brazil's Amazonian basin, where they found areas of incredibly rich soil they called terra preta. These may have been intentionally created or simply the result of dumping food wastes, bones, manure, and the remnants of cooking fires into a garbage pit. Either way, this combination of compost and charcoal is what created terra preta, soil that remains rich to this day. The benefits of biochar seem to last for millennia.

You could consider biochar to be humus on steroids; it works its magic by being extremely absorbent and having a huge amount of surface — a single gram can have a surface area of over 1,000 square yards. In addition to increasing the soil's CEC by absorbing and holding more nutrients in the soil, it also:

o Increases soil water-holding capacity by as much as 20%

o Hosts soil microbes, especially fungi, which help soil regenerate itself; mycorrhizae in particular seem to thrive around biochar

o Persists in soil, unlike most organic matter, continuing to maintain and improve soil fertility; biochar in the Amazon has been found to be thousands of years old

o Helps prevent disease

Although there is solid, promising research on the ability of biochar to suppress diseases, both soilborne and airborne, the mechanism by which biochar disrupts pathogens isn't clear. Most of the current theories attribute this effect to its improving the population of beneficial microbes, absorbing substances that damage roots, and/or generally improving the health of plants by enriching their environment. So biochar remains somewhat mysterious, but what *is* clear is that it benefits soil. It can be purchased at some farm supply stores or online, or you can make it at home in a couple of ways.

HOMEMADE BIOCHAR

The simplest way to make biochar is to burn scrap wood in the open. You can burn the wood in a trench or on the surface; a trench will create a ready-made planting bed once you've added soil and compost, or it can become the base of a compost pile. Biochar under a compost pile will get filled up with composted goodness — nutrients and microbes — that will live on and multiply when you mix it into your soil. The beauty of biochar is that nothing goes to waste; what can't be composted is turned into soil-enriching charcoal.

If you have a small backyard where burning a brush pile isn't appropriate, make your fire in a small metal drum or purchased fire pit. Obviously, you want to do this when there's no danger of wildfire; keep your material covered and dry, and burn it after a good rain. And follow your town's or state's open-burning regulations; some parts of the country require permits.

You can burn many different materials, such as woodworking scraps, corncobs and cornstalks, bamboo, and larger brush you've trimmed from the garden and let dry for a while. Incorporate smaller bits near the center of the pile so they don't fall off, and save some small, dry tinder for starting the fire on top.

Have a hose with a nozzle ready to go next to the fire. This is for safety and to prevent the wood from turning to ash. A metal rake to corral stray coals will come in handy.

1 Stack wood in a grid to allow for good airflow. Start with the largest wood on the bottom and leave a space about the size of the wood between pieces. Make a base the size you want.

2 Make the next layer the same way, but with pieces placed at a 90-degree angle to the first. Continue making layers like this until you've used up all your material, going from largest on the bottom to smallest on top — sort of the opposite of what you learned in your scouting troop. You'll end up with a fairly vertical pile. Break up the small brush you're left with and place it on top as tinder.

3 Now light the *top* of the pile. It's important to light the top because it creates a clean, virtually smokeless flame and makes biochar with very little ash. Use newspaper or a propane torch to get the tinder going; it should take off from there. You'll see a small amount of smoke until the fire gets going, at which point the smoke created by new wood being ignited below the fire is burned off by the flame.

4 Rake errant coals back into the center as the fire burns.

An upside-down fire creates less smoke and more charcoal.
Don't tell your scoutmaster.

Install Ventilation, Plumbing & a Nice Pantry

5 When everything is glowing but not flaming, douse the whole thing with water. This preserves the wood in the biochar phase and prevents it from turning into ash. It also rinses off whatever ash has formed — you don't generally want ashes in the garden unless you really want to raise the pH, and it's easy to overshoot with this amendment. Leave the fire to cool for a day or so.

6 Shovel the biochar into a wheelbarrow. (Fun note: finished biochar will make a tinkling sound when you stir it.) You may want to break it up a bit; a variety of chunk sizes — from powder up to ¾-inch pieces — seems to be ideal for improving soil texture and hosting a greater variety of organisms. To break up larger chunks, screw a small piece of plywood to a piece of 2×2 and pound on the pile with the end.

A fancier way to make biochar, which is highly efficient and smokeless, is to construct a biochar retort kiln. You'll need one 50-gallon metal drum and a 30-gallon one, both with lids, and a 3-foot piece of stovepipe. You can find detailed instructions on YouTube by searching for "biochar burner double barrel." This requires a tool (or a friend) that can cut some holes and slices in the metal drums.

Like other carbonaceous materials, biochar can tie up available nitrogen when used in the garden. The best remedy is to mix it with compost or something else that contains nitrogen. Biochar also soaks up a lot of water. So that you don't steal moisture away from neighboring plants, wet the biochar before you apply it unless you expect rain or plan on a good watering just after application.

To wet and mix biochar with compost, you can fill a 5-gallon bucket about three-quarters full with biochar, mix in a quart or two of compost or composted manure, then fill the bucket with water. If you're not too squeamish, you can also add urine, a great and quickly absorbed source of nitrogen (although it does need to be diluted). Stir the whole mess together, layer it on a planting bed, and mix it in lightly with a spading fork. You don't want to turn the soil — just stir it up a little; leave the mycorrhizae and other soil residents in relative peace as you feed them. This layer will be mixed in as you make planting holes. For extra points, throw in a handful of this biochar/compost mix every time you plant something.

So that you don't steal moisture away from neighboring plants, wet the biochar before you apply it unless you expect rain or plan on a good watering just after application.

This biochar is soaking up water in a bucket, instead of in your garden.

CHAPTER 5
FEED
THE INHABITANTS

Photosynthesis, Minerals & Soil Testing

In this chapter, we'll talk about the nutrients plants need; these nutrients include sugars made by photosynthesis and minerals, which the soil food web can generally provide. If your garden is floundering, though, you may want to do a soil test to find out which nutrients are missing. You can then give the soil food web a head start on balancing soil minerals.

I'll start with my personal favorite among chemical reactions: photosynthesis, the miraculous process that allows every living thing to be fed — directly or indirectly — by sunlight, air, and water, thereby making life on Earth possible. You see why I like it.

Photosynthesis: Empty but Necessary Calories

The most basic need for plants is the fuel they get from sugars. They make these from water, air, and sunshine via photosynthesis. They then turn those ingredients into life as we know it.

Sugars for plants are like our calories; without sugars, plants starve. The sugar produced by photosynthesis is used for many things:

o Sugars make plant structures. Plants literally build themselves by binding together long chains of glucose to make cellulose, which is what forms the "bones" of the plant.

o Sugar goes into substances the plant excretes or exudes, like nectar and root exudates. These feed the soil food web and other beneficial organisms, such as pollinators.

o Sugar ends up in things we eat, like fruits and vegetables.

Photosynthesis: The atomic shuffle, powered by sunlight

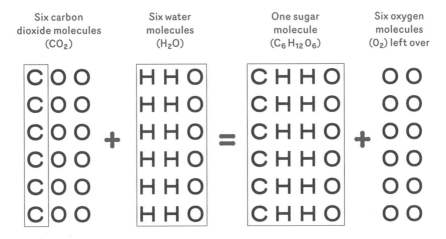

Where Does That Sugar Come From?

The sun works with chlorophyll (the stuff that colors leaves green) to create energy for plant construction and metabolism. When chlorophyll absorbs sunlight, one of its electrons gets excited and wiggly, setting into motion the chemical reaction that is photosynthesis.

Photosynthesis performs several critical functions. It:

- Pulls carbon dioxide from the air, mitigating carbon's greenhouse effect
- Sinks that carbon into the soil as sugars that feed microbes, or humus that improves soil texture while housing plant nutrients and soil microbes
- Turns carbon dioxide and water into oxygen and sugar (see diagram, opposite)
- Feeds those sugars to the plants that feed us and, like, everything

Think about it. Plants take carbon dioxide and water and, using photosynthesis, turn them into what we eat and breathe. Thanks, photosynthesis.

The sugars from photosynthesis provide the energy plants need. But they're like empty calories — alone, they would be akin to humans eating only white sugar. Sugars offer energy and building materials, but they don't have the mineral nutrients plants need. Those come from soil — specifically, from the soil food web.

Macronutrients & Micronutrients

The list of plant nutrients is pretty long, and no doubt there are other key nutrients we don't know about yet, because that's how science rolls. Here's what we know about macronutrients ("macro" because they use a lot of them) and micronutrients ("micro" because they use the teensiest bit) that plants need.

There are six macronutrients. The most noteworthy are:

- Nitrogen
- Phosphorus
- Potassium

These three are seen as the N-P-K numbers on fertilizer boxes. They've earned a place on the fertilizer label because plants use so much of them, and less-than-optimal soil is likely to be deficient in one or more.

The lesser of the macronutrients, meaning they're less likely to need replenishing, are:

- Calcium
- Magnesium
- Sulfur

Soil is less likely to be deficient in these elements, but a soil test will tell you for sure. More about that shortly.

The happy news is that adding organic matter, especially from a variety of sources, will help replenish all of these. You may be able to put that fertilizer bag back on the shelf.

This is a big subject, so let's go into a bit more detail about key nutrients.

The Big Three Macronutrients: Nitrogen, Phosphorus & Potassium

NITROGEN

Of the macronutrients, nitrogen could be considered the most important because it's vital to plant growth. Plants need a steady and generous supply, but it leaches (washes) or evaporates out of soil unless it's "immobilized" within soil organisms. This happens when something eats it or absorbs it from the air and uses it to build proteins for its own structure.

Nitrogen is an important component of chlorophyll, which is needed for photosynthesis. Chlorophyll makes leaves green, so a lack of nitrogen can cause yellowish leaves. Plants that are hungry for nitrogen are also slow growing and stunted.

Although 68 percent of air is nitrogen, plants can't generally access this. The exception is nitrogen-fixing plants such as legumes, which pull nitrogen from the air and deposit it in little nodules on their roots. Generally, though, plants need to get their nitrogen from the soil. In an ideal world, the soil food web takes care of this.

The best-case scenario for nitrogen is that members of the soil food web manufacture it on the spot — right in the root zone, where they like to live. Many soil microbes can absorb nitrogen from the air, making it available to plants. And nitrogen becomes a part of any organism that eats plant matter or other organisms. Because atoms don't just go away, and living organisms contain nitrogen, that nitrogen is either excreted (pooped out) or released when an organism dies. In healthy, lively soil, all those organisms are making more baby organisms, all of which contain nitrogen.

Using compost, mulch, and compost tea will get your soil on the road to nitrogen liveliness. Sometimes, though, that road is too long, and the gardener doesn't want to wait. If you're that gardener, you can add nitrogen, but use organic versions of it. Chemical (manufactured) fertilizers contain concentrated minerals and salts that will kill or chase away earthworms and their soil food web buddies. Your fertilizer should have an "N" rating of 10 or under, and that nitrogen should come in its original form (such as feather, blood, or fish meal); the labels on organic products list the ingredients.

Some good nitrogen-rich soil amendments are high-protein plant and animal sources (nitrogen is a component of protein). These forms include:

- Manure
- Blood meal
- Fish meal
- Soy meal
- Alfalfa meal or pellets
- Seed meals

If your soil is depleted or just crummy to begin with (as in my hydraulic-mined and then graded yard), you'll want to give it some nitrogen at first. Once your soil becomes rich in organic matter and soil organisms, it will feed itself, as long as you keep feeding the soil food web.

PHOSPHORUS

Like nitrogen, phosphorus plays a role in photosynthesis. It's also crucial to root development, strength and size of plant structure, and formation of flowers and seeds. Unlike nitrogen, phosphorus doesn't travel much — it stays within about an inch of where it starts out, so plants need to have phosphorus in the root zone.

When phosphorus is deficient, you may notice stunted growth and purplish or bluish discoloration of leaves. Because plants send needed nutients to newer growth first, you're likely to see symptoms of phosphorus deficiency on older leaves. A serious deficiency, however, will also interfere with flower, seed, and fruit production.

Most soils contain adequate phosphorus, but it's often tightly bound to clay, iron, or aluminum in the soil and therefore inaccessible to plants. Also, soil pH that's outside the ideal 6 to 7 range can keep phosphorus and other nutrients locked up. Nitrogen also figures into the picture: inadequate nitrogen will decrease plants' phosphorus uptake.

The correct pH adjustment makes phosphorus more available, and adding organic matter breaks clay up into soil aggregates or crumbs, allowing microbes to turn phosphorus into a form plants can use. The soil food web helps plants access phosphorus, too. Soil organisms, attracted by sugary exudates in the root zone, will adjust pH to that plant's liking, thus unlocking phosphorus. Fungi have an amazing ability to transport phosphorus to plant roots; mycorrhizae, the fungal strands that intertwine with roots and extend their reach, can bring nutrients from relatively great distances.

Adding chemical fertilizers that are high in phosphorus, however, can kill the very mycorrhizae that bring phosphorus to the root zone. And unless pH, nitrogen, soil aggregates, and soil organisms are at proper levels, any phosphorus you add to soil will almost immediately be locked up and inaccessible.

If your soil is degraded or otherwise starved, however, you can mix in a slow-release source of phosphorus. These include:

- Rock phosphate
- Bat guano
- Bonemeal, including fish bonemeal
- Granite dust (also a source of very slow-releasing trace minerals)
- Manure

Of these, I like manure best, especially cow or horse manure because they also contain nitrogen, potassium, and lots of organic matter in the form of partially digested hay and straw bedding.

As always, the soil food web is the hero here. No matter how much phosphorus you add to soil, it's unlikely to be in the root zone and digestible to the plant unless your soil contains mycorrhizae and other fungi. This is why organic matter is so important; it feeds the microbes that feed your plants.

POTASSIUM

Potassium is also involved in photosynthesis. It regulates the functioning of stomata, the openings in leaves that allow plants to take in carbon dioxide. And it regulates production of energy for chemical reactions, enzymes necessary for growth, and water circulation. Potassium can help plants be more drought resistant.

Plants deficient in potassium may have yellowing at the margins of older leaves. They may be stunted and more prone to wilting in dry conditions, and may lose leaves prematurely.

Good sources of potassium are:

o Kelp (added to the soil or used as a foliar spray)

o Wood ashes (use only on acidic soil, as ash raises pH)

o Greensand (a good slow-release, long-term source of potassium)

o Granite dust

Random side note: Bananas are a high-potassium food, both for humans and plants. Eat a banana every day and throw the peel in your compost. Your plants and your circulatory system will love you for it.

The Less-Famous Macronutrients: Calcium, Magnesium & Sulfur

CALCIUM

Adequate calcium in soil improves its texture and buffers pH. Calcium helps make cellulose, which gives plants their rigidity, and helps form roots, stems, and new growth.

Calcium deficiency is most visible in new growth, which may be stunted, deformed, or brown around the edges. You may also see yellowing between the veins of leaves. Insufficient calcium can affect bulb and fruit formation. Blossom-end rot in tomatoes, for example, is usually the result of inadequate calcium.

Calcium isn't mobile in plants once it's used for building structure, so plants need a steady supply. This is provided via transpiration, the circulatory system of plants that depends on the uptake of water through the roots. Because new growth transpires (evaporates) less water, it's more susceptible to calcium deficiency.

What you use to supplement soil calcium depends on your soil's pH and magnesium levels. A soil test will give you these measurements.

The ideal pH range for plants is between 6 and 7. If the pH is below the ideal, lime will both add calcium and raise the pH. Be aware that there are different types of lime, however. One, dolomitic lime, also contains magnesium, and an excess of magnesium in soil can lead to compaction and poor drainage.

If soil magnesium is adequate and the pH is low, use calcium carbonate to add calcium and raise the pH. Using dolomitic lime would add magnesium you don't need.

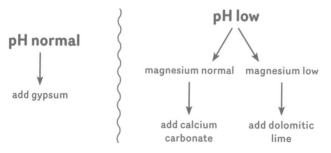

My soil needs calcium. What do I do?

pH normal → add gypsum

pH low:
- magnesium normal → add calcium carbonate
- magnesium low → add dolomitic lime

If your soil's pH is about right, use gypsum to add calcium. Gypsum contains calcium and sulfur; sulfur is another necessary macronutrient.

And if your brain is now about to explode, keep calm and see the algorithm on the previous page for deciding how to correct a calcium deficiency. Fixing a deficiency of calcium will also improve soil texture by helping to form aggregates (crumbs). That's why people add gypsum to heavy clay soil.

MAGNESIUM

Magnesium is necessary to the formation of chlorophyll. It also figures into carbohydrate metabolism. Magnesium can be leached from soil by water. An excess of potassium can also make it unavailable to plants. Soils that are rich in organic matter are unlikely to be low in magnesium.

A deficiency of magnesium shows up in plant leaves. Unlike calcium, magnesium stays mobile in plants; if there's not enough of it, the plant sends what magnesium it has to feed new growth. Therefore, signs of deficiency are most visible in old growth, primarily in yellowing between veins and at edges and, later, purplish, reddish, or brownish discoloration.

Adding organic matter is the best way to address magnesium deficiency, since it provides so many other benefits to soil. Dolomitic lime, sometimes used to correct acid soil, also adds magnesium; this can cause problems if magnesium is already sufficient. An excess of magnesium can cause soil compaction as well as a deficiency of calcium and potassium.

SULFUR

Sulfur is necessary to the production of chlorophyll and figures into the conversion of nitrogen to protein and the production of many enzymes necessary to plant metabolism.

Deficiencies can happen in sandy soil that drains too quickly, as sulfur is soluble and leaches from soil. Sulfur is used to treat too-high levels of many other minerals. Having too much sulfur is rare; it isn't recognized as an issue in agriculture. Fertilizers used to lower pH (acidify soil) also contain sulfur, as does gypsum, which is pH neutral.

The best way to add sulfur is by adding organic matter. (Do you notice a pattern here?) Organic matter adds sulfur while also bringing all the other benefits organic matter brings, which you should know about by now, if you've been paying attention.

Micronutrients

Sometimes called trace elements, these nutrients are needed by plants in much smaller quantities than macronutrients. There are between 6 and 13 micronutrients, depending on whom you ask. The 13 contenders are:

- Boron
- Chlorine
- Cobalt
- Copper
- Fluorine
- Iodine
- Iron
- Manganese
- Molybdenum
- Selenium
- Silicon
- Sodium
- Zinc

Your soil is unlikely to be deficient in any micronutrients, and organic matter from a variety of sources will supply these if they are lacking. And a mineral additive called Azomite contains a good balance of micronutrients; you can mix that into your compost for good measure. However, it's important to ensure your soil does not have a micronutrient imbalance. (For more information, search "signs of nutrient deficiency" at rsc.org.) What can happen, especially with iron, is that soil conditions — mainly the wrong pH — make micronutrients unavailable to plants. If, however, you have a robust soil food web, it will adjust pH and other factors that let plants access what they need from the soil. The soil food web's amazing like that.

How Plants Access Nutrients

Plants pull minerals from soil (and occasionally from the air, as in the case of nitrogen-fixing legumes). There are three main ways plants access mineral nutrients:

- **Transpiration or mass flow.** This is a simple mechanical process. As water is "exhaled" by leaves (this happens constantly, which is why plants wilt), it creates a vacuum. This pulls water from the root zone, along with whatever nutrients that water contains. Think of plants as being made up of tiny straws (it's not true, but it's a useful analogy). Those imaginary straws are the plant's circulatory system.

- **Osmosis or diffusion.** When nutrients in soil water are more concentrated than those in the plant roots, osmosis wants to even out the concentration. Nutrients move into the roots until the concentration is equal. Another example of osmosis is when salt pulls water out of the air to make the salt less concentrated. That's why it takes longer for your clothes to dry after you fall in the ocean than after you fall into a lake.
- **Root interception or direct contact.** Roots grow until they bump into soil colloids (humus or clay, usually) that contain nutrients, which they then absorb.

The catch here is that nutrients have to be in a form plants can absorb. This is where organic matter and the soil food web come in.

To Test or Not to Test

A soil test will provide many useful pieces of information, including readings of your soil's cation exchange capacity (CEC; the ability of soil to store mineral nutrients), pH values, amount of key macronutrients, percentage of organic matter, and more. If you're starting with soil that seems less than ideal, it's a good idea to have a soil test done. How do you recognize if soil is ideal? Healthy soil will have a lush, vigorous mix of weeds and grasses. When that soil is moist, you'll be able to easily pull up a handful of the plants growing in it. Underneath those plants, you'll find soft soil with lots of worms.

If your garden mainly features nitrogen-fixing weeds like clover and vetch, and nitrogen-fixing trees like locust, your soil is likely nitrogen deficient. These leguminous plants make their own nitrogen, but plants that can't fix nitrogen will limp along in deficient soil. Other deficiencies cause different kinds of limps (some that we talked about in the macronutrients section above), but the presence and health of existing plants will give you an idea of your soil condition. If there's just not much growing, you know your soil isn't supporting much life.

Bottom line: if your soil or the plants growing in it seem unhappy, test it.

How to Test

To test your soil, purchase a bag from a soil testing lab (you can find local labs online) and dig up some soil (either from one spot or from several to give a composite reading). Let the soil dry out, crumble it into small pieces and mix it up, then put the specified amount in the bag and mail it to the lab. Depending on how much you want to spend, you could pay for separate tests for garden areas that have different growing needs — a lawn and a vegetable garden, for instance.

Prices will depend on the extensiveness of the test you select. Some tests go beyond the usual pH, macronutrients, and CEC readings to examine soil for micronutrient levels, ECe (a measure of soil salinity or "saltiness"), and the presence of heavy metals.

A couple of weeks after you send in your bag o' dirt, you'll get an e-mail with an incomprehensible graph showing the results of your soil test, along with a nearly incomprehensible umpteen-page explanation. Here's what you actually need to know.

What Test Results Mean

Here are the most common components of a soil test, with the ideal ranges and how to get there.

PH — ACIDIC OR ALKALINE?

GOAL: 6 to 6.8, usually

The pH scale ranges from 1 to 14. A pH of 7 is neutral. Lower than 7 indicates acidic soil, and higher than 7 is alkaline. Like the Richter scale used to measure earthquakes, the pH scale is logarithmic, so the differences between numbers can be misleading. The further you get from neutral, the more it takes to change the number. So a soil with a pH of 6 is 10 times more acidic than a soil with a pH of 7. Most plants thrive with a pH between 6 and 6.8; if you shoot for 6.5, you're usually about right.

Interestingly, soil pH varies seasonally, and plant preferences coincide with those variations. Cool-season vegetables and leafy greens prefer more alkaline soil, which happens during cooler temperatures when biological activity slows. Fruiting plants like tomatoes, melons, cucumbers, and squash want more acidic soil, which occurs with the booming soil food web action of summer. Like many things in gardening, nature tends to get it right without our help.

Soil organic matter helps compensate for pH that's out of whack by creating a more-ideal pH in the root zone. The sugary substances that plants exude from their roots attract soil organisms that are ideal for that plant, meaning they produce the correct pH and mix of nutrients. As always, the best practice is to add organic matter.

Sometimes, though, the pH is too far from ideal and we need to tweak it for what we want to grow. On my own property, the soil that wasn't scraped off by grading is under pine and oak trees, where the litter of pine needles and leaves tends to produce a fungally dominated, acidic soil. In those areas, I add compost and oyster shell lime to gently shift the pH. Much of my lot, however, was graded, leaving me with subsoil that's alkaline. This is gradually being corrected by my use of compost and mulch.

If your pH is too low (too acidic), add calcium; see the algorithm on page 92 to determine which kind of calcium to use. If your pH is too high (alkaline), add nitrogen, which lowers pH by feeding microbial activity, or sulfur, which feeds a particular type of bacteria that makes sulfuric acid; elemental sulfur is considered an organic amendment. Sulfuric acid sounds gnarly, but it is natural. Sulfur, however, isn't a quick fix — it requires weeks or months and the presence of the proper microbes to work. Your approach to correcting soil problems should always be multifaceted, and it should encourage and feed microbial growth. Do this by adding mulch, compost, and compost tea (more about this in chapter 6).

 Soil organic matter helps compensate for pH that's out of whack by creating a more-ideal pH in the root zone.

ORGANIC MATTER

GOAL: 4 to 10%

The sky is *almost* the limit on this one. It's unlikely you'll ever have too much organic matter, but if you've added lots of uncomposted matter and don't have enough biological activity to digest it, that situation is possible. This can tie up nutrients until the soil food web breaks the organic matter down into nutrients and humus. High organic matter accompanied by a lackluster CEC means your organic matter isn't benefiting your soil; the results are considered together.

WHAT TO DO:

- **High organic matter, CEC >20.** Smile in a self-satisfied way and keep doing what you're doing.

- **High organic matter, CEC <20.** Add life, in the form of compost and/or compost tea, to break down organic matter into something that holds on to nutrients.

- **Medium organic matter.** Practice standard maintenance — use compost and mulch to maintain organic matter and nourish microbial activity.

- **Low organic matter.** Add cover crops, lots of compost, mulch, and compost tea to increase microbial activity. If soil is compacted, you can dig or till in compost *once*. After that, tilling does more harm than good by impeding the biological activity you've added.

NITROGEN

GOAL: 20 to 30 ppm

Nitrogen is needed in substantial amounts, but it is quickly used by plants and can leach or evaporate from soil; the most stable form is stored in humus or microbial bodies. Adequate nitrogen is necessary to plant growth and green foliage, indicating the presence of chlorophyll and the ability to photosynthesize.

There can be too much of this good thing. Excess nitrogen can cause growth that is too rapid and weak, making plants more susceptible to disease; increase water usage; and even cause nitrate poisoning in you or animals that eat plant greens.

Nitrogen will eventually be supplied in adequate amounts by a healthy soil food web, as biological activity produces it. If your soil test shows it's way off, however, you'll want to take corrective measures.

WHAT TO DO:

- **High nitrogen.** Plant heavy feeders like tomatoes, corn, cucurbits (squash, cucumbers), cabbage, and grass cover crops.

- **Medium nitrogen.** Keep up the good work.

- **Low nitrogen.** Add rich compost and a balanced organic fertilizer if phosphorus and potassium are also low or medium, or a nitrogen-rich fertilizer (like blood meal or fish meal) if phosphorus and potassium are high.

- **Very low nitrogen with medium organic matter.** Add blood meal or fish meal.

- **Very low nitrogen with high organic matter.** Add a slow-release source like feather meal, since all that organic matter will be releasing nitrogen, too — sort of a ticking nitrogen bomb. If your plants look anemic during the growing season, you can use a foliar spray of nitrogen, such as fish emulsion; this will be taken up quickly and will leach away by the time the organic matter kicks in.

HELPFUL RULE

If your soil test gives you per-acre recommendations for amendments, use this equation:

pounds/acre divided by 440 = pounds/100 square feet (a 10 × 10-foot area)

PHOSPHORUS P1 OR WEAK BRAY

GOAL OF WEAK BRAY: 25 to 40 ppm

The "weak bray" phosphorus reading is what you'll want to act on, as this shows available phosphorus.

WHAT TO DO:

- **High or medium phosphorus.** Do nothing. Because phosphorus is mostly locked up with calcium, high levels don't do any harm; the phosphorus will just hang around until biological activity releases the perfect amount to plants. Adding phosphorus, however, can increase pH because it's usually combined with calcium, an antacid like the one in Tums.

- **Low phosphorus.** Add soft rock phosphate, bonemeal, or fish bone meal. Be sure your soil is biologically active, as phosphorus is unavailable to plants until soil organisms break it down.

POTASSIUM

GOAL: 60 to 100 ppm

Potassium is important to plant metabolism, strength, and vigor, but too-high potassium interferes with the uptake of other nutrients and raises soil pH.

WHAT TO DO:

- **High potassium.** Add gypsum. Don't use combination fertilizers, which always include some potassium (the "K" in N-P-K).

- **Low potassium.** Add compost. First eat a lot of fruits and vegetables, especially bananas, and compost the peels. Add kelp meal, greens, and/or wood ashes in moderation (wood ashes can increase lime and pH more than you'd want).

MAGNESIUM

GOAL: 50 to 150 ppm

Magnesium is important in plant metabolism, but an excess can bind soil particles together too tightly, making it difficult for water, air, and nutrients to circulate. Low magnesium can occur in soils low in organic matter or in places with heavy rains.

WHAT TO DO:

- **High magnesium.** Add gypsum.
- **Low magnesium.** Add compost, which is rich in magnesium. Lime also contains magnesium but will raise the pH in the bargain.

CALCIUM

Calcium is important for plant cell walls. Too much can raise pH, causing soil to be overly alkaline. Too little can cause blossom-end rot and other plant woes. What you use to add calcium will depend on several other factors; see the algorithm on page 92 to decide which kind of calcium you add.

WHAT TO DO:

- **High calcium.** Add compost or, if you're in a hurry (very high pH, for instance), sulfur.
- **Low calcium.** See page 92.

SODIUM

Sodium, a type of salt, can cause mischief by raising pH and interfering with nutrient uptake. High salinity causes osmosis (the process by which roots absorb water) to go backward, basically sucking the life out of roots.

WHAT TO DO:

- **High sodium.** Add gypsum. Water more deeply and less frequently. You may need to test your water to see if it's high in sodium or other salts (see ECe or Electrical Conductivity, page 102).
- **Low sodium.** No such thing. There's always enough, which isn't much.

ECe OR ELECTRICAL CONDUCTIVITY

ECe measures soil's ability to conduct electricity; this is basically a test for salinity from excess sodium, or from high levels of other salts. If ECe is high but sodium is normal, have your water tested for high levels of salts such as calcium bicarbonate, magnesium carbonate, potassium nitrate, chlorides, and sulphates.

Deep irrigation can "rinse" salts out of soil, provided the water itself isn't high in salts. Shallow, frequent irrigation, especially with water that contains salts, allows water to evaporate near the surface and leave salts behind. Water less frequently and more deeply, and mulch well to reduce evaporation and thus the need to water.

If your irrigation water is highly saline, you may want to buy a filter; these are pricey, however. You might have to live with lower yields. Or move.

SULFUR

GOAL: ≥20 ppm

Sulfur is most commonly used to lower pH, but it's also an important nutrient for plants.

WHAT TO DO:

- **High sulfur.** No such thing, pretty much. Sulfur has to be extremely high to cause problems.

- **Low sulfur and high pH.** Add elemental sulfur (an organic amendment that lowers pH).

- **Low sulfur and good pH (6.0-6.8).** Add gypsum, which is pH neutral.

MICRONUTRIENTS

Micronutrients (sometimes called trace elements) shown in soil tests include zinc, manganese, copper, iron, and boron. Of these, boron is most likely to cause problems if it's too high. Other problems with trace minerals are generally caused by an imbalance among them, or by pH that's out of range. Correcting pH and adding Azomite will fix most micronutrient issues.

WHAT TO DO:

- **High boron.** Add gypsum.

- **Any low trace minerals.** Add Azomite, a mineral-rich rock mined in Utah from an ancient volcanic bed. Use granules, not dust, to provide a slow release of trace minerals, since plants only need tiny amounts of them.

PERCENT CATION SATURATION

This measures the balance of cations, which are positively charged ions. The primary cations in soil are potassium, magnesium, calcium, and sodium. The focus of this test is the proportions of each — the balance between them.

The ideal percentages of these cations is:

- Potassium 4 to 7%

- Magnesium 10 to 20%

- Calcium 65 to 75%

- Sodium <3%

Because calcium can tie up other nutrients, you'll want to bring it up to the ideal range before you try to correct other things (see previous information about how to do this with specific nutrients).

CATION EXCHANGE CAPACITY (CEC)
GOAL: >20 ppm
We covered this earlier; see pages 77–78.

EXCESS LIME
GOAL: not "H"
A high reading here is probably caused by the application of lime. If you've been adding lime, quit it. If your calcium is high, add sulfur to bring it and your probably high pH down.

What to Do with the Results

If your soil test is not wildly deranged (with results well outside ideal), the best approach is to correct deficiencies by using good soil husbandry: add organic matter and increase biological activity by adding mulch, compost, and compost tea. These practices will eventually correct almost all soil problems and will benefit your soil more than adding amendments.

Organic matter makes your soil more resilient and allows the soil food web to adjust at a comfortable rate; you want to keep your soil food web happy by not throwing it too many curve balls in the form of drastic changes. Mass infusions of nutrients can cause overgrowth of the microbes that eat them, throwing off the delicate balance of things that live in soil. That balance is what helps keep diseases and pests in check and creates the ideal environment for each plant.

If, however, results are very low or very high, or if pH, sodium, lime, or calcium are very high, you'll probably want to use amendments to correct things more quickly, in addition to adding organic matter. Results at either extreme of the scale indicate problems that can dramatically affect plants' ability to thrive in your soil.

If you decide to use amendments (additives) to correct imbalances, make corrections gently and in the right order. Always use organic sources of minerals; their effect is less drastic and more long-acting, and they coexist well with the soil food web, which is damaged by chemical fertilizers. Natural sources also contain a mixture of macro- and micronutrients and trace elements; it's like getting your vitamins from food instead of a pill.

ORDER OF OPERATIONS FOR FIXING SOIL

1 CORRECT PH. You already know how important pH is. If the pH is too low, add calcium; if it's too high, add sulfur.

2 FIX YOUR CALCIUM "PERCENT CATION SATURATION." Calcium figures into the uptake of other nutrients, cell division, and disease suppression; it also lowers nitrogen needs. Adequate calcium loosens the soil and makes nitrogen more available; too much can tie up other nutrients.

Calcium binds with many other soil nutrients, so you'll want to get your calcium up to at least 60 percent of total cation saturation first, before you try to add other nutrients; otherwise the new calcium will just keep soaking up nutrients you've added. Adequate calcium needs to be your baseline.

3 CORRECT EXCESSES. Most excesses are corrected using gypsum, a combination of calcium and sulfur. If your calcium is already high, you can still use gypsum, as sulfur is the correction for high calcium. Sulfur leaches out excess minerals.

4 CORRECT MACRONUTRIENT DEFICIENCIES. Nitrogen, phosphorus, and potassium (the N-P-K numbers on fertilizer labels) are emphasized because they are used in large quantities and the mobilized forms leach out readily in water or evaporate. See individual nutrients on pages 88–91 for how to correct.

5 CORRECT MICRONUTRIENT AND TRACE ELEMENT DEFICIENCIES. Most problems with micronutrients are caused by an imbalance among them. Azomite, an ancient seabed deposit from Utah, contains micronutrients and trace elements.

Chemical N-P-K fertilizer is bullying some adorable little microbes.

CHAPTER 6
COMPOST
& COMPOST TEA

It's Not Rocket Science

Once you've built the house (healthy soil), you want the inhabitants to live there happily, so you need to feed them. The soil food web is designed to feed plants, but I like to give it a little boost by adding more friends and growing food for it. This is what compost is good for. Your soil food family loves to dine on decomposing organic matter (and each other); compost is a concentrated source of what they prefer to eat.

For new gardens with poor soil, you'll want to mix in compost and any other organic material you've got, to get some life in the soil and improve its texture. You might even need to rototill. *Once.* And for the record, the mulch you're continually throwing down will eventually be composted in place by the soil food web. Adding compost from your pile is sort of like serving microbial guests at the dining table, as opposed to making them get up and go to the buffet.

Benefits and Tools

Compost isn't complicated, and it does many wonderful things for your soil:

- It improves structure by helping form soil aggregates or crumbs.
- It adds nutrients in plant-available form.
- It adds soil organisms — those things that turn garbage into compost and go on to continue improving the soil.
- It adds diversity, so no one bad guy can dominate.

And it's not rocket science. There are only five, maybe six, things you need for a successful compost pile:

1 GREEN AND BROWN ORGANIC MATTER.

2 A HOSE and easy access to water.

3 A SMALL AMOUNT OF REAL ESTATE (about 4 to 6 feet square), or even better, enough space for two piles so you can flip the pile back and forth, but this isn't necessary.

4 A SPADING FORK or pitchfork for turning the pile.

5 SOME CLIPPERS to cut up long pieces of brush or vines. Keep these clippers next to the compost pile so you won't be tempted to just toss in long pieces, which will bring you sorrow down the road.

6 BONUS ITEM: a covered bale of straw or other dry material in case the compost gets too wet — usually from rain. You can even use shredded sensitive documents; the compost will digest away any harmful chemicals in the paper, and your documents will be so thoroughly shredded that even a criminal armed with methamphetamines and Scotch tape won't be able to piece them back together. Fallen, dry duff works, too, if you have some you can rake onto the pile.

That's it!

SOIL IMPROVEMENT, ONE PLANT AT A TIME

While you're making global improvements in your soil, you can also improve it each time you plant something. Plants themselves will improve soil by putting out exudates that attract nutrient-producing organisms. Compound this effect by adding nutrients each time you set something in the ground.

Whenever you plant seeds or seedlings, bring along a bucket of compost. For seeds, pull up or move aside any cover crop or mulch, chop a row with a hand hoe, then layer on compost and chop it in. Next, plant the seeds, stir them in with the hoe, and pat the soil down over them. These rows are ideal for smaller crops like carrots and beets.

For seedlings and larger seeds like corn, beans, and nasturtiums, pull up the cover crop in a small area, use the hoe to scoot aside the mulch, and chop in a little compost. Then make a hole the right depth for that seed or seedling, plant it, and pat. (Hint: Nasturtiums look great mixed in between vegetables, and they repel pests.)

Using this method, you inoculate your soil with compost and worms every time you plant something. That compost goes to work feeding what you've planted and spreading organisms to compost the mulch you've put there. And the plants do what plants do: attract nutrients to their root zones and build soil.

Is It Compost or Is It Mulch?

Well, either. Or both. Basically, mulch is organic matter that sits on top of the soil. And compost is the finished product after you put the same stuff in a pile and help it rot until it looks like dirt and less like . . . stuff.

If you leave mulch on top of the soil, it will eventually decompose into compost. You've probably noticed the rich, dark layer that develops under pine needles or leaves: that's compost.

Mulch and compost serve similar, important functions (improving soil texture, encouraging microbial life, and providing food for worms and germs). They just do it on a different timetable. Think of compost as sort of a sourdough starter (a healthy mix of worms and germs) and mulch as the flour (food for that starter). Both work together to break up hard, dead dirt clods and turn them into the kind of rich crumbly soil that brings tears of joy to the eyes of gardeners. And plants, if they had eyes.

Both mulch and compost have their charms. Compost is rich in the microbial life that will eventually break down mulch. Mulch protects compost by keeping the soil surface moist and hospitable to worms, while slowly turning into compost itself.

Compost can, of course, be used as mulch, and often this is all you need to do to maintain good soil. As long as it doesn't dry out too much, the good organisms will trickle down. Top-dressing with compost once your soil is healthy is often a better idea than digging it in, because digging disturbs the fragile and vital mycorrhizae. But keep the compost moist. Cover it with mulch, and water the surface now and then.

 Think of compost as sort of a sourdough starter and mulch as the flour.

How to Build a Compost Pile

Step 1: Pick the Right Spot

Find a spot that is near a hose with a nozzle. Water needs to be easily and permanently available — you don't want to have to drag a hose over to wet your compost. You *know* you're not gonna do it.

Ideally, the compost spot should also have two adjacent, compost pile-sized areas so you can turn the pile to and fro every week or so. Turning a compost pile completely while moistening it mixes textures and microbes, encourages microbial growth, and aerates it to keep stinky anaerobes at bay.

Step 2: Start Piling

Use whatever organic matter (anything that is currently or formerly alive) you have or can find. You'll need two kinds:

"**GREENS.**" These are things that add nitrogen and rot quickly:

- **Kitchen waste.** Vegetable peelings, apple cores, that limp celery that's been in the "crisper" since you used a few stalks of it for stuffing last Thanksgiving

- **Yard waste.** Grass clippings, leaves you've raked up (or, preferably, that your neighbor raked up and thoughtfully bagged for you), prunings from shrubs and trees, spent plants from your vegetable garden

- **Some types of manure.** Manure from farm animals (dog and cat manure has too many microbes that can make humans sick)

"BROWNS." This is stuff that's brown and dry:

o Straw

o Newspapers

o Paper from your shredder (paper comes from trees, which used to be alive)

o Dead leaves, pine needles, or most any other organic matter that has fallen on the ground

You're shooting for a mix, about three parts by volume of brown, dead matter to one part of fresh green stuff. The greens feed microbes and other members of the soil food web. Browns won't provide as much food for microbes, but they will give structure to your compost, keeping it from becoming a compacted, stinking mess. Woodier brown materials also contain more lignin, the hard structural cells that turn into humus.

None of the ingredients should be much more than 4 inches long. Longer pieces create a tangled mess that's difficult to turn. Smaller pieces are easier to turn and compost faster; you can chop kitchen waste and yard trimmings into 1-inch pieces to further accelerate the process.

Layer whatever you have to start with in a little (or big) heap, mixing up the ingredients.

Step 3: Water It

Water each layer as you go. It should feel as moist as a wrung-out washcloth — damp but not wet on your skin. It shouldn't leave a wet spot on you when you touch it. Keeping the level of moisture right is key. Too wet, and the pile becomes stinky. Too dry, and it sits there and refuses to rot.

Step 4: Turn It

Turn the pile once after you build the layers, to make sure everything is nicely mixed. Check the moisture level when you do this, and correct it, either with a squirt of the hose or by adding dry material — straw, dry leaves, paper from your shredder, or torn-up newspaper.

COMPOST PANACEA

Turning your pile, while correcting moisture, is the most important thing you can do for your compost pile. This gives good germs — the real workhorses of compost — what they need to go forth and multiply. When they do that, they turn raw materials into humus and nutrient-rich compost. To quote a song from the Great Folk Scare of the '60s: "turn, turn, turn."

Maintaining a Compost Pile: It Needs Your Love

Every week or two, turn the pile by picking up forkfuls and flipping them over onto the adjacent compost pile spot. When you do this, "correct" the moisture level to the wrung-out washcloth level. If it's too wet, from rain or too much gloppy kitchen waste, add dry browns. If it's too dry, sprinkle each layer with a hose as you turn it. Use the fork to toss the compost up and down as you let it drop onto the new spot; this helps break up and aerate any chunks of stuck-together green waste.

Note: If you fail at compost husbandry, you'll still eventually have compost. Weather and time will decompose anything on the ground. The process won't bring you joy the way an active compost pile does, but it will manage to rot without your help.

Adding to Your Compost

Save food waste in a bucket or adorable container of some kind on your kitchen counter (I use an enameled ice bucket, which makes kitchen waste look good while it's waiting for me to take it out). Along with fruit and vegetable scraps and coffee grounds, you can include *small* bits of meat or eggs. The more variety, the better. If you want compost soon, chop up corncobs, broccoli stems, and other hard, thick pieces of vegetable waste.

You can also throw in anything alive or formerly alive from your yard that, for whatever reason, you don't want to chop up and leave in place. This can include spent garden plants, prunings, and things you rake up. As always, cut up long pieces so they'll get cozy, rot, and be easy to turn.

I don't worry about excluding diseased plants or weed seeds. I think adding disease organisms to compost gives their enemies something to practice on, so they can combat them later (this parallels the human immune system and is the reason babies put everything

COMPOST BINS & OTHER CRUEL HOAXES

You *can* buy one of those fancy compost makers, but people tend to pile stuff in them and ignore it. Six months later, when they're planting vegetables and want some nice fresh compost, they find the contents are either dry and dead or wet and stinky. Bins make it unlikely you'll ever turn your compost. Compost needs love, not a container.

If you do use a container, make sure it's one you can lift up, set next to the pile, and turn the pile into. This can be as simple as fence wire made into a circle, or it can be a purchased container designed to be used that way.

Compost tumblers (rotating barrels) require regular turning and attention to moisture levels. Also, being out of contact with the ground, they can't attract worms, microbes, and other things that live in soil and make for good compost; you'll have to add those. And all containers are pricey.

The advantage of the rotating compost tumbler is that it keeps out vermin; rats enjoy a good compost pile (the square-sided plastic containers won't keep them out; a wire container works better, if you relocate and turn the pile regularly to be sure they haven't burrowed under). It's a good idea to keep a rat trap near your compost pile to let you know if rats have discovered it, and to catch them before they move in.

WHAT NOT TO ADD TO YOUR COMPOST

- Pet manure; this can contain germs that make people sick.
- Big chunks of meat scraps.
- Long weeds, branches, or vines; they make compost impossible to turn. You're making a compost pile, not a brush heap.
- Live Bermuda grass. Bermuda grass is the spawn of Satan. It will regrow wherever you put it, and if you chop it up, it will turn into little Bermuda grass plants. Lay Bermuda grass out in the sun with roots exposed for a week or so, until it's good and dead, before you compost it.
- Lots of people won't add diseased plants or prunings to their compost pile, but I do. See the Frequently Asked Questions, Answered section (page 123) for more about that.

KEEPING VARMINTS OUT

If you don't have a cat, or your cat is lazy and useless, set a couple of rat traps near the pile. If you have skunks, raccoons, or dogs digging in the pile, keep a big container of hot red pepper flakes and sprinkle some on each time you add things or turn the pile. Raccoons and other animals are one reason you might want to use a container. Containers on the ground won't keep animals out, but an elevated rotating one will.

in their mouths). And if weed seeds sprout, they'll be easy to remove from your soft, mulched ground. As you putter, you'll pull them up and throw them down as part of the mulch.

The miracle of biodiversity is that it's self-correcting. If you build a good compost pile, it will turn into good compost that, when you offer it to a plant, will provide what that plant needs. As with all things garden, diversity is divine.

When Is It Done?

Compost is finished when you can no longer identify what it was made of. It looks like dirt. *Really good* dirt. Be sure not to let compost dry out before you use it in the garden. The important thing about compost is its biological activity, which needs moisture to live.

If you want compost in a hurry, start with smaller pieces and turn and moisten it every day to get those microbes excited. Give the pile some *fine*, protein-rich food to get the germs to work; this can be grass clippings, soy meal, oatmeal, or the alfalfa pellets you feed rabbits. Otherwise, don't add anything new. The microbes won't let you down — you can have usable compost in as little as a week. And if you need a few handfuls of compost for something you're planting, just scoot the pile over and grab some of the rich, brown dirt at the bottom. You can "dilute" this with moist potting soil, peat moss, and/or perlite to make it go further. Any worms that come along for the ride are a bonus.

Top secret hint: Like much of life (art, writing, being a grown-up), there are degrees of "doneness." I have no prohibition against burrowing under my compost at any stage and throwing a few handfuls of that rich, brown stuff, as long as it smells good, into a watering can. I then add water from my "water feature" (otherwise known as a cow trough with goldfish in it) and use this wonderfully germy mixture to water plants that need extra love. You can also use it like any compost, mixing it into planting holes. There just won't be as much of it as from a finished compost pile.

Growing Pains

As with childrearing (many parallels here), making compost can take you through some rough phases. Like children, your compost can sometimes be obstinate, smelly, or inert. Usually, though, it will shape up with time and love.

How do you fix a compost pile that isn't living up to its potential (either not rotting or rotting in a way that smells nasty)? Here's what to do for each difficult phase.

RECALCITRANCE. If your compost just won't rot, add more small, green stuff — grass clippings, alfalfa pellets (rabbit food) or meal, shredded green branches, or kitchen waste cut into small pieces. You can even put your kitchen waste into the blender and pour the resulting slop over the compost pile. Smaller bits have more surface area for microbes to feed on, and microbes make compost.

Then be sure the pile is as moist as a wrung-out washcloth. Too little moisture is the most common reason compost doesn't . . . compost.

 **Like children, your compost can sometimes be obstinate, smelly, or inert. Usually, though, it will shape up with time and love.**

SMELLINESS. If your compost is stinky, turn it to introduce air, and add more brown matter — straw, dry leaves, shredded paper, chipped dead branches, duff you've raked up from the environs.

Turn it often, correcting the moisture level each time. You'll soon be thrilled and amazed — the bad smell will be gone, worms will be frolicking, birdies will be singing. Life will be good.

Depending on many variables, your compost may be, by turns, cold, hot, wet and smelly, dry and lifeless, or just right. The solution is pretty much always the same: turn it and correct the moisture level.

Fungally or Bacterially Dominated?

Although all compost is good for all soil, certain types give more of a boost to different plants. In general, fast-growing, short-lived plants such as vegetables and annual flowers prefer bacterially dominated, higher-nitrogen compost, and slower-growing plants (perennials, shrubs, and trees) like their compost fungally dominated. I think of the former as living fast and dying young, imbibing lots of bacteria/nitrogen on the way.

Interestingly, the type of plant you want to feed makes the kind of compost it wants to be fed. This makes sense when you think about it. Browner, slower-growing plants drop dry leaves and branches, mulching themselves with organic matter that's friendly to fungi. And fast-growing plants die and create a greener, fresher mulch. Nature's got this. We try to replicate what nature does with our compost (and mulch, which is basically future compost).

To make a more fungally dominated compost, use a higher proportion of browns. This will also keep your compost cooler, which is more conducive to fungal growth.

For a bacterially dominated compost, add lots of green, nitrogen-rich material: grass clippings, kitchen waste, fresh plant waste, and chipped green prunings. Smaller pieces have more surface area and feed more bacteria. Bacteria are what heats up a compost pile, so you'll know you've got lots of them when your compost gets warm.

SIDE NOTE: pH AND COMPOST

pH is a measure of how acidic or alkali something is. A pH of 7 is neutral, neither acidic nor alkaline. Most garden plants like their pH a little on the acidic side, maybe 6 to 6.8.

Long-lived plants such as trees and shrubs prefer fungally dominated soil, which is more acidic. This is facilitated by larger pieces of brown organic matter on top of the soil — kind of what these plants drop. More than coincidence, eh, nature?

Short-lived plants, like vegetables and flowers, prefer bacterially dominated soil, which is more alkali. This is facilitated by smaller pieces of green organic matter, dug into the soil.

For a bacterially dominated compost, add lots of green, nitrogen-rich material: grass clippings, kitchen waste, fresh plant waste, and chipped green prunings.

Worm Bins

Worm bins are a great invention for managing kitchen wastes while growing worms and creating great, casting-rich compost ("worm castings" is a nicer name for "worm poop").

Worm castings are considered the best fertilizer there is, and they're not hard to get. Good compost will have lots of worms and their castings, but a worm bin has only one purpose: to feed worms. A worm bin will manage kitchen waste from most households, without smelling bad, and while making great compost.

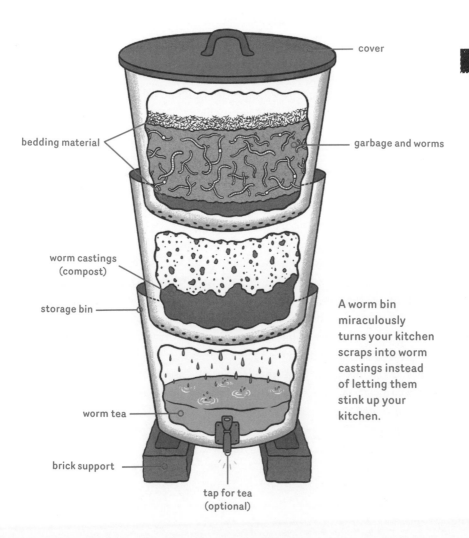

A worm bin miraculously turns your kitchen scraps into worm castings instead of letting them stink up your kitchen.

There are lots of designs for worm bins. The book *Worms Eat My Garbage* is a great resource for more information, but I'll give you a distillation here.

Basically, you'll need a container that allows air in and liquid out. The more surface area you have, the better worms like it and the more food you can add. If you add kitchen waste faster than the worms can eat it, your worm bin will become anaerobic and smelly. If the proportion is just right, the worms will eat the waste and aerate the soil, keeping the whole mess fragrant, in an earthy way. This proportion has a lot to do with the population of worms, so you can add food faster when the population is greater.

The container can be a wide, shallow wooden box, a plastic bin with holes in the sides, or a commercial worm farm — a series of boxes with mesh bottoms that stack and are digested from the bottom up. All of these containers need a spigot of some sort and something waterproof at the bottom to protect your floor or deck. Worms will also need some bedding that isn't green, such as shredded newspaper or other paper, coir, dry leaves, or small wood chips.

Making a Bin

To set up a bin, first put in a good layer of moist bedding material 3 to 4 inches thick (moist, shredded newspapers or coir are good). Add some soil and compost that contains red wigglers, or buy worms commercially, by mail or from certain garden supply stores. Drop in a little more bedding, then start adding kitchen wastes for the worms to eat, burying the food in the bedding to discourage bad smells and fruit flies. This should be a very shallow layer, and you don't want to add more until the worms have eaten most of it and increased in numbers. If your worm bin is crawling with worms, you can toss in more bedding and mix in food faster, but be careful not to let kitchen wastes pile up to the point that they're rotting instead of being eaten by worms. Obviously the amount of food scraps will depend on how concentrated the worm population is.

Always put in an equal amount of bedding after you add new kitchen waste; all those new worms need someplace to live.

From time to time, you can drain off leachate, otherwise known as worm tea. This makes a great fertilizer. Mix about one part of tea to at least four parts of water and use it to water plants you love or that need your love. Commercial worm bins have a spigot on the bottom tray for this purpose. You can tip other bins onto their corners and drain whatever liquid is at the bottom into the tray. Harvest this leachate often, before it gets anaerobic and smells bad. If you're itching for some worm tea and nothing drains, you can water the bin and drain what settles in the bottom; just be sure and drain it well. You don't want a soggy bottom.

Frequently Asked Questions, Answered

》》》 **Q: Why isn't the compost pile heating up? Shouldn't it be hot?**

A: Many composters think that success equals hot and steamy. This may be true of romance novels, but compost is another story, with a more nuanced definition of success.

Hot compost's claim to fame is that it kills weed seeds and pathogens. In our world, however, we don't try to kill anything in the garden; we add diversity. Cold compost contains a bigger variety of microbes — that is, diversity. It's also more hospitable to worms.

The heat in hot compost comes from microbial digestion of nitrogen-rich contents, like grass clippings, kitchen waste, or green yard waste. If there's a high proportion of greens to browns, the compost will heat up. I'll admit that a steaming compost pile makes me happy. Compost's job, however, is not to make me happy, but to grow microbes for the garden.

Cold compost, one type of which is vermiculture (for worms, or *verms*), rots at a lower temperature. It's true it doesn't kill weed seeds and pathogens. It also doesn't kill a whole bunch of other microbes; if aerated, those microbes will be overwhelmingly beneficial. And your deep mulch will make it easy to pull out any seeds that sprout.

Cold compost is more conducive to growth of fungi, which require and produce less heat. Fungi are vital to a thriving ecosystem; they decompose brown matter that bacteria don't address so much. Plants will take what they need from the mix of microbes you offer them.

The main advantage of hot compost is that it breaks down more quickly. If I'm in a hurry and want to heat up a pile, I'll add green matter and turn it more often. You can mix in manure if you have it (livestock manure, not pet), but this isn't necessary. Alfalfa meal or pellets or soy meal will also provide heat-producing nitrogen, as will plain old green plant matter.

Personally, I don't worry about the temperature of my compost, so long as it's aerated and smells good. I just want aerobic life to put into my garden.

>>> Q: What about diseases?

A: I do not fear diseases in my compost. A disease microbe is not one that came from outer space and landed on your peach tree; it's just a microbe from the neighborhood making itself a little too much at home, like a party guest that stays late and eats all the guacamole. (For more thoughts on this, see the column by Mike McGrath, Howard Garrett, and famed gardener and pruner Lee Reich in the bibliography.)

Don't kill that bad microbe, whose name is Legion; introduce it to a large variety of better-behaved ones who will "eat or compete": keep it in check, either by eating what it eats or eating *it*. (A party guest is perhaps not the best analogy here . . .)

The vast majority of garden microbes are beneficial or neutral, and you'll want a lot of them around. Adding plants damaged by pathogenic microbes — ones that cause disease — to compost gives the nice microbes a chance to practice fighting them. It also infuses compost with those competing germs, so you can introduce them to the garden.

>>> Q: What about weed seeds?

A: I don't fear weed seeds either. When I effortlessly pull those sprouted seeds out of my soft, mulched soil and throw them down, they add organic matter, keep the soil moist, and feed the soil food web.

⟫⟫ Q: Why does my compost smell bad?

A: Usually too much moisture and not enough air. An overload of kitchen waste sometimes does this if you don't "dilute" it with browns. Turn your pile and add dry stuff (dry leaves, shredded paper, straw, pine needles). On a sunny day, you can often just rake dry duff onto the pile from nearby. Turn the pile again in a couple of days. It will be better. I promise.

⟫⟫ Q: Why is my compost just sitting there? Shouldn't it be doing something?

A: If your compost is just sitting around looking bored, it's probably because it's either too dry or it lacks green, nitrogen-rich material that heats up with bacteria. Compost needs moisture and nitrogen to rot.

You need to keep the pile moist. That's why I told you to locate it near a hose and source of water, but did you listen? *No.* Turn and sprinkle the pile if it's too dry; dry compost is dead compost.

Once your compost is moist, be sure it has nitrogen-rich food for the bacteria. Some sources of nitrogen are:

o Green garden waste, or kitchen waste

o Alfalfa meal or pellets

o Soy meal

o Feather meal

o Manure

o Cottonseed meal is also rich in nitrogen, but it's a pesticide-intensive crop; avoid unless it's labeled organic

If you have large branches, cut them up so they'll lie on the ground in contact with dirt and other compost bits. Branches that are grouped together to form little "tents" hold too much air and let the compost dry out (think brush pile). Microbes need to be moist.

In summary, to resuscitate dry, bored, boring compost:

- Turn on a hose with a nozzle. Be sure it's a nozzle with a trigger-style shutoff, so the hose doesn't do that flailing-about-spraying-water-all-over-and-terrifying-the-cat thing.
- Spray the top with water liberally, then turn the pile, one forkful at a time, to the adjacent spot, spritzing each layer again as you move it. If there are branchy pieces, cut them to about 4 inches, and be sure they lie flat — no forks sticking out in different directions.
- Sprinkle some small green or otherwise nitrogen-rich stuff on each layer; this can be fresh grass clippings, kitchen or yard waste, manure, or alfalfa meal or pellets. This is bacteria food, and bacteria get compost going.
- Check the moisture level; make sure it feels damp but doesn't leave water on you when you touch it. If you're a really dedicated compost rescuer, turn it again, correcting the moisture as you go. Rake some duff from around the pile on top to shelter it, and sprinkle that.
- Now, step back and watch the pile explode with worms and germs. Figuratively speaking.

After a day or two, the compost will be revived. Turn it again for extra credit and enjoyment.

⟩⟩⟩ Q: What are those disgusting giant-maggoty-looking things?

A: Black soldier fly larvae. They're a good thing; they can manage a compost pile that's become smelly and anaerobic from too much kitchen waste, transforming it into food for worms. And chickens love the larvae. See opposite for more information — they're kind of amazing.

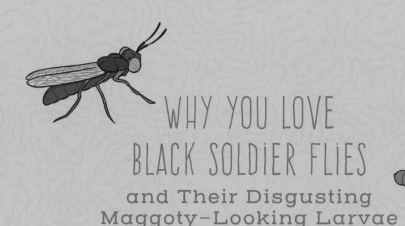

WHY YOU LOVE BLACK SOLDIER FLIES
and Their Disgusting Maggoty-Looking Larvae

Black soldier fly (*Hermetia illucens*) is a common and widespread fly that is all good, despite its maggoty-looking larvae.

Black soldier flies do only two things: (1) make maggots to break down your less-than-stellar compost and (2) hatch into flies that look a little like mud-dauber wasps but don't sting or bite. They don't want to do anything but mate and make more little disgusting maggoty things.

Those maggots do many wonderful things. They:

- Work with worms to break down compost.

- Compete with other flies we don't really love, like houseflies and biting flies.

- Sanitize garbage, turning rotting food into something nicer that worms like.

- Break down harmful chemicals into harmless compounds.

- Decrease E. coli and salmonella populations in chickens.

- Fix stinky compost by breaking it down and generating heat, which helps dry it out.

When you find black soldier fly larvae in your compost, give them love. And turn your compost more often, so it's not stinky and rotten, which is what attracts soldier flies.

Compost Tea for a Non-Rocket Scientist with a Bucket

Compost tea takes the goodness of compost — nutrients and, mainly, microbes — and makes it easy to spread around. Pour it or spray it; one advantage of being able to spray those microbes is that you can inoculate a lot of soil area quickly, with not much compost. Once you have a good variety of organic matter in the form of mulch, the microbes in compost tea will jump in and start feeding on that mulch, creating a bustling metropolis of organisms and breaking down the mulch into new compost.

Another big advantage is that you can spread microbes onto plant branches and foliage. Like human skin, plant surfaces are crawling with microbes, most of which are beneficial, and all of which are kept in check by living with a large number and variety of other microbes. Different conditions — like damp growing areas with poor light and air circulation, or past chemical use — can damage this community of organisms, known as a microbiome, and favor disease-causing microbes. Tea from healthy, aerobic compost adds back organisms that compete with the pathogens.

Why Compost Tea?

When pathogens come a-callin', trying to kill them is both futile and counterproductive. Those microbes are everywhere; using chemicals to zap them also kills other microbes in the area, most of which are beneficial. Then the pathogens will come back to an impoverished microbiome where they can thrive unimpeded. If we instead increase the number and variety of microbes, those will eat or compete with the disease-causing ones and lessen their effect.

As you know, if you've been paying attention, pathogenic organisms tend to be anaerobic — they don't need and often can't tolerate oxygen — and beneficial ones tend to be aerobic, thriving in well-ventilated places. One way to get a healthy population of aerobic microbes is to make compost tea with a good air supply, or "actively aerated compost tea." This sounds like a big deal, but it's really a pretty simple process.

Here's what you'll need to make actively aerated compost tea:

o An aquarium pump with two outlets

o About 2 feet of ¼-inch "leaky hose" irrigation tubing (if you can't find this, just use a couple of air stones for aquariums, which may or may not be included with the pump)

o One 5-gallon bucket

o A couple of pieces of duct tape, maybe 6 inches long each (a brick will also work)

o Powdered kelp and molasses, for germ food (optional)

o One or two quarts of compost that's alive and fragrant; this quantity is approximate — the important thing is that it's lively compost

A compost tea-making device is actually pretty simple. It's basically a bucket and an aquarium pump.

Since germs stick to things with their slime (remember your teeth in the morning), you want to bubble, or aerate, your compost tea, so as to knock the germs off into the tea. Aeration also encourages aerobic, not anaerobic, microbes. That's why this tea is called actively aerated.

How to Make Compost Tea

1 To the ends of the plastic tubing that comes with the aquarium pump, attach the "leaky hose" irrigation tubing. It'll fit on the adapters that are supposed to go to the air stone for your fish tank. Honest. If it doesn't come with adapters, you can use drip irrigation connectors for ¼-inch hose. Or just use two aquarium air stones.

2 Curl the leaky hose in the bottom of the bucket and tape it to the bottom in a couple of places with duct tape, to keep it from floating to the top. You can also hold it down with a brick, if you find yourself out of duct tape. But who runs out of duct tape?

3 Fill the bucket three-quarters full of water and bubble it for a few hours to remove chlorine. Toss in a handful of kelp and about 1 tablespoon of molasses if you have them, but these are frosting on the microbial cake.

4 Throw in several handfuls of good compost. By "good," I mean fairly well rotted, alive, fragrant, and moist — as opposed to dried up and dead. Ideally, this should be enough to cover the bottom and be a couple of inches deep. If you fill up an empty ½-gallon milk carton, that'll be about right.

5 Let the tea bubble at room temperature for 24 to 48 hours. Protect it from sunlight, which kills germs. It will form a brown, scummy foam. This is a good sign that microbes are flourishing. Keep it bubbling until you're ready to use it, but no more than 2 days.

As always, the smell is key. Bad germs smell bad; your compost tea should smell like loam, or a clean lake. If it stinks, pour it in an out-of-the-way part of your garden and start over.

When it's ready, pour off the liquid into a watering can and use it to drench the plants you love the best. Or you can strain it into a garden sprayer: scoop it up with a mixing bowl with a pouring spout, then pour it into the sprayer through a tea strainer. If the strainer gets clogged, simply tap the flotsam back out into the bucket.

Using a sprayer is a little more work, but it lets you inoculate many more plants. I do this when I want to prevent or treat foliar disease, or just to give a little boost to a lot of plants.

Apply compost tea when the temperature is mild and there's no sunlight hitting the places you spray. A cloudy day or dewy morning is a good time. You want the microbes to be able to populate the surfaces before they dry out.

Spray the undersides of leaves to give the microbes a sheltered spot to take hold. They'll fan out to eat or compete with the bad guys, who don't like biodiversity. You can also use a sprayer to distribute good germs to your soil. Ideally, do this when it's raining or while watering. The germs will soak in through your mulch and infuse the soil with life.

I've been experimenting with adding leaves from diseased plants to an already-robust compost tea. The idea is to feed the organisms that eat the disease microbes, causing a little orgy of beneficials. I did this with peach leaf curl, then sprayed the inoculated tea on some diseased trees. So far the new growth is disease-free. I will need to repeat this experiment to prove that the compost tea caused the improvement, and not the dryer weather or other factors. But it's undeniably exciting. To a compost nerd, anyway.

 Apply compost tea when the temperature is mild and there's no sunlight hitting the places you spray.

Building a Garden That Feeds Itself

I've talked to lots of people who claim they have a "black thumb" (i.e., a thumb of death versus a green thumb). For the record, someone with a black thumb is usually just somebody without a water timer and snail bait. Or a duck. (Ducks love snails.) Follow my simple instructions and change that thumb color. Green is the new black.

Once you have a few tools — including a sprinkler, a hose, and a water timer — you're halfway there. Pick a spot you can water, make simple paths (because we hate hardpan), pull stuff up, and throw stuff down. You've just loosened your soil and created mulch, aka future compost. Soon you'll be ready to plant!

Step 1: Pick a Spot

First, select an area that's manageable; 4 × 8 feet is a nice size to start. Do you want a little flower garden or vegetable plot (or — my preference — a combination)? Pick a location that you'll visit often and that you can easily get water to.

Although it's nice to choose a spot with good soil, it's not necessary, since you'll be improving it. My own front garden, which now stops traffic (a bicyclist yesterday said, "This is the prettiest garden in the county!"), was initially graded gray clay, with broken asphalt dumped there by the county road crews for good measure. The practices outlined in the book have turned it into fertile soil.

 To have robust, living soil, you need to keep it moist.

Step 2: Get Water to It

The soil food web doesn't like to dry out. Some organisms will die, and others will just go dormant and pout. To have robust, living soil, you need to keep it moist.

Be sure the watering system is set up before you begin. For the record, hand-watering doesn't work, generally. It takes more time for water to penetrate to deep roots than people are inclined to spend when watering by hand. Get a sprinkler and a timer that goes on the faucet (a cheap kitchen timer–type shutoff will do). Run it for about 15 minutes, then dig up a trowelful of dirt to see how far the water penetrated. Shoot for penetration about a foot deep; adjust your sprinkler time accordingly. Then you just turn on the sprinkler the first time you go out for the day and again in the evening if you have seeds or seedlings. The garden will be watered, and you won't discover you forgot to turn it off a couple of days later, when the neighbors are wondering why their wells went dry. I kind of might have actually done this once.

Once your soil is deeply moistened and you've mulched the area, water less often to encourage plant roots to dig for water. Wait a day or two, dig down to check moisture, and adjust your watering schedule to keep the moisture about right. Of course, new seeds have to be kept moist and thus will need more attention. Be sure their root areas are mulched.

Step 3: Decide Where You'll Walk

If your planting area is less than 4 feet wide and you can reach into it from all sides, you don't need to build paths. If it's bigger, though, figure out where you'll walk and plan to make permanent paths. That way you won't compact the soil, crushing members of the soil food web and creating hardpan, which is hard for roots to penetrate. For a refresher on how to make paths, see chapter 4.

Step 4: Soften the Soil

The ground needs to be softened before you start gardening in it; this requires a couple of good soakings, either by rain or by the sprinkler you've put there. If your plot isn't already overgrown, let it get that way so you have something to grab when you start to pull up weeds (the next step). My favorite time to begin a garden plot is late fall or early winter, when the ground is soft from rain and I can let everything lie fallow until I plant. But you can do it any time.

Give the area a few days or weeks of water and neglect; more won't hurt. If you want to, pick up a bale of straw (wheat or rice; organic is best) or alfalfa hay while you wait for the area to soften and grow. You can also beg the neighbors for bags of leaves or yard waste — this is a good precedent to set, because your yard will always benefit from more mulch.

Wait a day or two after a good soaking, so the ground is deeply moist and soft but not soggy. This is critical; if you work your soil when it's too wet, it will be gummy and won't get loosened. Too dry, and the weeds will break off from the roots. When the moisture is just right, weeds will pull out along with their roots and surrounding soil, and clumps of dirt will break up into crumbs.

When the time is right, put on your play clothes and prepare to get dirty. If you're blazing new paths, bring a bucket or wheelbarrowful of wood chips.

Step 5: Pull Stuff Up, Throw Stuff Down

Sporting old clothes and garden gloves, go into your future garden and pull out anything growing there — it will generally be a combination of weeds and grasses. Shake the rich dirt off the roots (plants improve soil, remember?) and onto the planting beds, then lay down the plants. You're making sheet compost. How easy was that?

Start by pulling up everything from the 2- or 3-foot-wide path you have planned, then gradually work your way in, pulling from the planting beds you can reach from the path. (Don't step on the beds!) Throw the plants onto the beds, then toss wood chips (or whatever material you choose) on the path as you clear it. This will create some visual order in your garden. Throw grass clippings on the bed if you have them.

Step 6: Mulch

Pile mulch on top of the pulled weeds. Use whatever you have — straw, leaves, cut-up prunings, even pine needles; the more diverse your mulch, the better. This layer of mulch will shade the pulled weeds so they can compost. It will also keep them from resprouting, and later it will rot and feed the soil. Any weeds that *do* sprout will be easy to pull out of the soft soil you're creating.

To ensure you have a thick layer of mulch to begin with, you may want to buy some. One effective combination is alfalfa hay — which is rich in nitrogen and protozoa — covered by straw, which shelters and nourishes whatever is under it. Layer this deeply (6 inches or so is great, but do what you can with what you have).

FUN FACT ABOUT NATURE, pH, AND PINE NEEDLES

If given a good mixture of organic material and microbes, nature will break down pine needles and other plant matter that is allelopathic (growth-suppressing) into compost that nourishes growth, *and* it will correct the pH to fit the plants that grow there. Nature's amazing like that.

Step 7: Plant Stuff

Now that you've prepared your planting bed, it's time to plant!

Planting Small Plants

When planting seeds, seedlings, and small plants — say, up to 4-inch pots (even gallon-size if the soil is fairly loose) — use a triangular hand hoe (like a Ken Ho) or a spading fork, depending on the size of the plant; never use a shovel. Chop a hole rather than dig one. Chop compost into the hole, improving the soil one plant at a time, making sure the hole's edges are jagged and extend for a little ways beyond the plant's roots. Pull out any clumps of plants or roots from the dirt you've loosened. If you pull out a weed, shake the dirt around the roots into the planting hole, and use the green part for mulch once you've finished planting. If it's a plant you like, move it to a new spot of your choosing. The end result is a rough-sided planting area with unimproved soil and compost loosely mixed, and surrounded by loosened, decreasingly improved soil.

This method inoculates the area with compost, worms, and microbes while encouraging roots and worms to venture into the surrounding unimproved soil. Then blanket the soil around the plant generously with compost and mulch. As the compost is worked into the loosened soil by worms, the entire area begins to improve.

If your soil is sticky clay, you'll want to add something to loosen it up; in gummy soil areas, you might throw in perlite and gypsum along with compost. (Gypsum breaks up clay soil while adding calcium and sulfur, both of which benefit foothills soil.) If you live in the forest and your soil isn't yet full of microbes (which can break down phosphorus for plants), you might want to add phosphate — a soil test will tell you if you need this, but it won't hurt. You can either use rock phosphate or a mix specific to your region from a local garden center. With time and the addition of organic matter, soil will correct itself, but using this mix will save you and your new garden some heartache.

In areas I've planted for a few years, it becomes increasingly rare to find a hard clump of my formerly dreadful soil. Worms, microbes, and roots have tilled it for me.

Planting Trees and Shrubberies

Remember hearing that when planting a tree or shrub, you should dig a hole three times as wide and as deep as your rootball? *Pfft*. From my experience, it's best to dig a hole with jagged sides (a spading fork works well) just large enough to hold the roots, fill the hole with the same soil you dug out of it, and cover the top with compost and mulch. These (relatively) impoverished circumstances lead to healthier, heartier trees and shrubs. Let me explain.

Those giant holes that diligent gardeners dig and then fill with amended soil are great for a while. The plant roots grow into the fabulously soft soil, quickly filling the planting hole and getting a little spoiled in this rarified, coddled environment. Then reality hits. The roots reach the real world, in the form of the sides of the hole (neatly dug with a shovel).

Like spoiled children faced with the difficult transition from the cushy life Mommy and Daddy have provided all their young lives to the hard, unamended, and (in my case) gummy clay soil of the real world, they make the only rational choice: "We're never leaving home." They turn around, go back inside where life is easy, pour a bowl of Cheerios, and get comfortable on the couch watching *Mister Rogers*. They invite all their little root friends to do the same. Sure, they're a little stunted, but life's easier inside.

Meanwhile, the plant starts to languish as its roots get more and more overcrowded and eat up all the food inside the hole. The soil inside is root-bound, and the soil outside remains unimproved by roots that might have invaded and softened it. A heavy rain or forgotten hose may fill this hole with water that stagnates, causing root rot and the death of the plant.

Contrast this with roots that are exposed to a hardscrabble life from the time you adopt them. You dig a hole just big enough to hold them, with rough sides. You fill in around the roots with the same unimproved soil they'll have to venture out into when they grow up, mixing in compost or soil only if necessary to allow soil to sift in around the roots. Because you were too lazy to dig the hole deep enough (this is a hypothetical situation, of course), some of the roots are still aboveground when you fill in the hole. You — quick! — find more crummy dirt to pile over the exposed roots, then, because you feel guilty for digging such a small hole, you mound lots of compost and mulch over the whole thing. For good measure, you loosen the soil in "rays" around the pathetic little hole you dug, inserting your spading fork and just lifting the soil a little, thinking this might give the roots somewhere to go outside their tiny, unimproved home. You pile compost and mulch over the area of rays, as well.

Then, a miracle happens. (Miracles are, by the way, commonplace when you let nature take its course.) The roots, after a brief initial period of indignation at the lack of coddling in their new accommodations, decide they had better figure out how to fend for themselves. They start to grow, putting out exudates that attract what they need from the soil and the compost above, from which worms and germs are beginning to trickle down. They pry apart clods of dirt, because there's not much else to choose from, and become accustomed to making their way in a tough world.

Then — whoa! — they begin to find greener pastures: softer "rays" of dirt that have been penetrated by nutrients and worms from the compost above. They grow into the jagged-sided rays and into the compost-amended topsoil and begin to transform their inhospitable world into a better one.

SOIL AMENDMENT, ONE PLANT AT A TIME

Plants, in concert with the soil food web, are all you need to create rich soil. As time goes on, I find myself squeezing in plants among other plants; empty spots become increasingly rare the longer I garden. This is good; all plant life improves soil. Plants feed the soil food web. And the soil food web feeds plants, increases organic matter, and adds biodiversity both above- and belowground, attracting more beneficial bugs and microbes and keeping pests in check.

In areas where I've been growing things for a few years, my soil becomes soft and rich. This happens not from any large-scale tilling in of organic matter, but because each time I add seeds or plants, I chop up a little more soil and add a handful of compost. I often add a mycorrhizal mix as well, which nourishes not only what I'm planting but other roots in the neighborhood. The plant roots then go on to attract nutrients, break up the soil, and add organic matter when they die or are pulled out. *Any* life is good for soil, pretty much.

Far from depleting it, packing your garden with a variety of plants will only enrich your soil; life attracts life. Your soil doesn't need fertilizer — it needs plants.

TOOLS YOU SHOULD LOVE & PERHAPS GET

KEN HO GARDEN WEEDER. This is my *very favorite* garden tool. The Ken Ho is a triangular hand hoe. It's great for weeding — you can slice weeds off at the surface, leaving the roots to improve the soil, or you can use the point to pick out small weeds. And you can use it to dig small to medium-sized planting holes or to lift up and transplant small plants.

A Ken Ho is my favorite tool. I own four, which allows three to be lost at any given time. Because that's how I roll.

FISKARS CLIPPERS AND FELCO HOLSTER. These clippers have a thin, sharp blade. They don't rust, so, in the unlikely event you happen to forget them and leave them in the rain for a week, they're not ruined — just open and close them under soapy water and spray them with WD-40, and they're good as new. I also like the lock being on the top instead of the side because it's easier to find on autopilot. I *always* carry clippers when in the garden so that it's easy to prune any errant branches while I putter, chop them up on the spot, and throw them onto my mulch. Tidier garden, happier worms. Felco makes a clipper holster you can hook on your belt or pocket.

RUBBER-DIPPED GLOVES. If I wear gloves, I don't have to sandblast my hands to get them presentable for work. The knit ones with rubber-dipped fingertips and palms keep my fingernails clean and make me brave in the face of spiders and stickers.

FISKARS MACHETE. A machete is *great* if you have major yard cleanup. You trim branches and chop them up in place. No trip to the dump, no dragging them over to a brush pile to burn; they turn into mulch, aka worm food. And it's exhilarating to wield a machete, especially if you're mad at somebody who, I'm sure, has earned it. The Fiskars model is well designed, has a plastic handle that doesn't give you splinters, and doesn't rust.

A WATERING SYSTEM. You can, of course, have someone put in a pricey irrigation system, but you don't have to. Get a sprinkler, a hose that reaches your garden spot, and a cheap kitchen timer–type water shutoff timer.

CHEAP GARDENING SOCKS. These keep your feet from being trashed, as gardeners' feet often are. I buy these by the six-pack and wear them inside cheap plastic clogs.

PUMICE STONE/NAIL BRUSH. This is for when, despite all your efforts, your hands or feet start to look disreputable.

GARDEN SHREDDER. This is the one big splurge I recommend you make, if you can afford it. As you cut berry canes, branches from trees you're pruning, or dead garden plants, you feed them into the shredder, which chips them into wonderful, rich mulch.

A shredder also makes short work of pests. My kale has at times been plagued by harlequin bugs and aphids. When I pull up the spent plants, I shred them, chopping up the insect eggs and remaining bugs. This damages pests and makes them vulnerable to disease — kind of like inoculating your garden with bug cholera. I then work the chips into the soil to feed beneficials that feed on the pests. Take THAT, harlequin bugs.

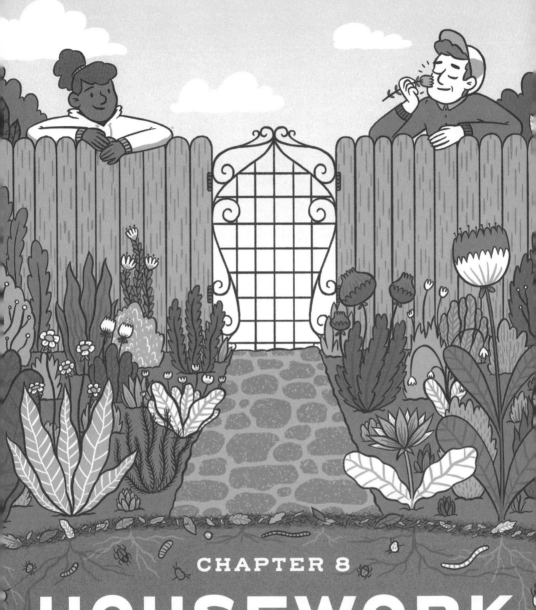

CHAPTER 8
HOUSEWORK
& HOME REPAIRS

How *Not* to Make Your Neighbors Hate You

Once you've created your beautiful, sustainable, diverse garden, your neighbors with manicured lawns will stop by and compliment you on it, while thinking, "Sure, it looks great now, but I know how you sustainable types are. Pretty soon this yard will be all overgrown and full of weeds and dead squash plants."

This can happen. It won't happen to you, though, because I'm going to give you some pointers on how to easily keep your garden looking loved and beautiful. The exciting part is that any tidying you do will be building soil.

Maintenance = Mulch

Maintenance doesn't have to be exhausting if you develop simple, ongoing habits, all made easier by the practice of keeping a thick mulch on your soil. Mulch keeps soil soft, making it easier to pull weeds. It also gives you a handy place to throw stuff you cut off or pull up; i.e., everywhere. All that garden waste serves to shelter and feed the soil.

Mulching should be a never-ending process. Whenever you're in the garden, toss on new organic matter: pulled weeds, raked-up leaves and pine needles, and prunings you've cut or broken into short (3- to 4-inch) pieces. The more quantity and diversity of mulch on your soil, the happier soil microbes will be.

Putter, Don't Work

Puttering — which includes deadheading, casual weeding, little nibbles of pruning, and, ideally, a cup of coffee — goes a long way toward keeping your garden tidy. My morning putter is often the best part of my day. I make discoveries in my ever-changing garden while also tidying it up. This is often how I find out that some berries are ripe, or some flower has started blooming.

Try to always carry pruning shears with you (in your own yard; not necessarily at work or in the airport) so you can trim spent flowers, dead stalks, and misplaced branches as you go. You can purchase a holster that makes it easy to clip the pruning shears to your waistband. Keep shears on the porch or by the front door to make it easy to grab them when you head outside.

If you putter often enough, you'll spend a lot less time on major yard cleanups because you'll be doing maintenance every time you walk through the garden! Using this approach, it's easy to keep your garden tidy even as you add more nutrients to your soil and feed the soil food web. When more serious pruning is required — as when growth in your garden has outpaced your puttering efforts — you can perform "machete maintenance" (see page 153), an effective and liberating approach to pruning.

 **Try to always carry pruning shears with you (in your own yard; not necessarily at work or in the airport) so you can trim spent flowers, dead stalks, and misplaced branches as you go.**

Let's look at the basic puttering principles for keeping a tidy garden, in order of slothfulness.

- **Puttering 101:** Weeding and its finer points. Pull stuff up, throw stuff down.

- **Puttering 102:** Deadheading, the best-kept secret. Snip and flip. Plus cutting off dead branches and anything ugly that's likely to stay that way.

- **Puttering 103:** Piecemeal pruning. Chop and drop. Maintenance as mulch.

Puttering 101: Weeding — Pull Stuff Up, Throw Stuff Down

While strolling in your garden and marveling at the wonders of nature, keep an eye out for sprouting weeds. Watch especially for sprouting grasses: they can make a garden look neglected in a hurry, but they make a rich addition to your mulch. Pull weeds up and throw them down; you'll find that it's gratifying to yank out something you don't like and use it to nourish something you adore.

If you pull up weeds when they're small and lay them under plants you love, for mulch, your garden will stay tidy and you'll conserve moisture and feed the soil food web. Small weeds are also easier to pull up. Left unchecked, though, little weeds turn into, well, big ugly weeds that make your garden look messy and spawn more baby weeds. Bigger weeds take a little more work to manage, but it can be done, and they'll also serve as mulch.

Caveat: be sure your soil is moist when you weed. When you try to pull weeds out of hard, dry dirt, they just break off at the surface while the roots stay behind, scoffing at you and plotting their next sprouting session.

TWO-FINGERED WEEDING

This is the method you'll use for small weeds that you catch early. When you see tiny weed seedlings growing among your garden plants, just daintily remove them with your thumb and index finger and drop them on the mulch.

Weed roots pull out more easily after a rain or a good watering, but if they're small enough and your mulch is deep enough, the soil moisture level won't much matter, because small roots haven't had a chance to grow through the mulch into the soil. If you habitually pull up weeds whenever you're out, they'll become increasingly rare in your garden, since they won't be making seeds.

TWO-FISTED WEEDING

Big weeds call for a stronger approach. I use this method when I'm breaking new ground for planting, or in weedy areas that have gotten away from me. Put on your gloves, grab fistfuls of weeds, then shake off the dirt and throw them down on the same spot, where they'll compost in place.

The mulch (aka weeds) you're laying down makes future weeds easy to pull up; this sets up a wonderful cycle. If a weed has become huge, just cut it into smaller pieces and scatter it to blend in with the mulch. Watering before weeding helps the roots to let go of the soil, but if roots don't come out with the weed, all is not lost. When you regularly remove the aboveground growth, roots eventually die and feed the soil. A triangular hand hoe (like a Ken Ho) is good for slicing off new growth at the surface.

If you're looking at a large weedy planting bed and thinking you'd really rather go inside and have a beer, don't despair. Make a deal with yourself that you'll spend 15 minutes on a small area (say, a square yard) and pull up *all* the weeds there. Then go inside and have a beer. I often find weeding addictive once I start and end up doing more. But you don't have to. Just commit to your 15 minutes. Then do another chunk the next time you're out.

WHAT ABOUT SEEDS & DISEASES?

I don't worry about throwing weed seeds down with my mulch for two reasons:

1 Those seeds are *frigging everywhere*. Dragging my weed seeds to the garbage can or burning them isn't going to keep the millions that are left from sprouting.

2 Throwing out the weed gone to seed is a waste of food for my soil. Weeds are "green manure." That plant, and the babies it's soon to make, will add mulch and nutrients to my soil. The deep mulch will make the seedlings easy to pull out. And a good variety of sprouted weeds ensures a good choice of nutrients for your plants. I think of those weed seedlings as the alfalfa sprouts of my worm 'n' germ salad bar.

If a particular weed has run amok or is especially hateful, target it for that season, looking for and removing it more zealously than the rest. Pull up the plant (ideally but not necessarily before it makes seeds) and toss it down. Throw the targeted weeds in a spot where they will be easy to spot, and pull any babies that sprout; this lets you complete its elimination during the next generation.

I've followed this approach with those horrible tiny burrs. I'm currently targeting dock, an unpleasant-looking weed that makes big, ugly, prolific seed heads. Dock resprouts enthusiastically from its deep, hard-to-pull-up roots, so I regularly pull or chop off any new growth so as to starve the roots. The new leaves get smaller and eventually stop growing, and the roots rot into organic matter in the soil.

Over several garden seasons, as long you don't turn the soil and bring buried seeds to light, you'll see fewer and fewer weeds. Birds and wind will still bring some in, but they become simple to manage.

Likewise, I don't worry about throwing down diseased clippings from plants. Removing them won't get rid of the pathogens, which are everywhere, but cutting and scattering these clippings will give competitors a chance to work on them and flourish.

So don't curse that weed — it's feeding your soil. Thank it for its service. Then yank it and leave it in the sun to wither and die.

Puttering 102: Deadheading — Snip and Flip

Deadheading deserves its own category. Perhaps the most important way to keep your garden from looking disreputable is to regularly remove ugly, dead stuff. Do this as you stroll, and scatter said dead stuff on the spot as mulch, soon to be compost; no bagging up prunings and hauling them to the compost pile or green waste bin. (BTW: "Waste" is a singularly appropriate name for that bin. All that plant matter should be feeding your garden.)

Focus first on getting rid of undesirables that make your garden look neglected in a hurry: dead flowers, dead leaves on veggies, things that have gone to seed. Simply snip and flip them, tossing them in different directions as you cut. (Or put them in a paper bag if you want to save the seeds to plant elsewhere. I often throw down chard and kale seeds in a new area, to grow another day.) Next, remove dead or ugly branches and break or cut them into 3- to 4-inch pieces, scattering the bits as you go. The idea is to make small pieces and spread them evenly around, so the clippings look like mulch instead of a bunch of dead stuff you just cut off and left on the ground, which, for the record, is what it is. And all that mulch makes the soil soft, rich, and easy to pull weeds from.

PERFECTIONISM, THE ENEMY OF MAINTENANCE

There are finer points to deadheading. If you ignore these, however, life will somehow continue, and your neighbors will like you better. That's because imperfect deadheading is more likely to get accomplished, and it rejuvenates plants and makes a far more presentable garden.

Here's one example of perfectionist deadheading: You *should* cut roses back to an outward-facing bud above a five-leaf stem. This bud will grow into a flowering stem. However, just breaking off dead flowers and tossing them down is quick and easy; if you wait until you can deadhead the "correct" way, your roses will look ratty and your perfectionist soul will be crushed. If you simply break off dead roses whenever you're out, your rosebushes will be just fine; they'll work around your human failings and continue to bloom. And your garden will look good. You can do more perfect pruning on an official cleanup day.

Puttering 103: Piecemeal Pruning — Chop and Drop

Pruning is a huge topic. My favorite books on the subject are by Cass Turnbull and Lee Reich. But if you don't want to go that deeply into the subject, you can approximate good pruning with my "one-third rule": now and then, when you're out with your pruning shears, remove about one-third of the branches of perennials that have finished blooming, or any shrub or tree that's getting too bushy and crowded. You can do this all at once or prune a little whenever you're puttering with clippers; just be sure not to remove more than one-third of branches in a given year. The exceptions are peach trees and nectarine trees, which like to be pruned more severely. Remove *two-thirds* of the branches from peach and nectarine trees and trim off two-thirds of the ones left, because these trees fruit on new growth.

How do you decide what to cut? Start with dead branches — the no-brainers. Then remove branches from areas that are too bushy or crowded. If you're not at the one-third mark yet, take off more; branches growing into the middle of the tree are good choices. In general, cut each branch you're removing back to the point where it originates; this keeps it from sprouting a bunch of new, crowd-y branches from the tip where you cut it.

FUN FACT ABOUT PRUNING TREES

To control the size of a tree, prune it in summer, when it's fully leafed out. You'll be limiting the nutrients those leaves would have made, and the tree will adjust to its new diet by reducing its size.

The one-third rule lets you sidestep the trickier aspects of pruning, like knowing whether a particular plant blooms on new or old wood; you'll be leaving some of each. It's a pretty safe way to prune, because you can't go too far wrong. The worst you can do is lose a third of older branches that would have bloomed or fruited.

Cut anything you prune into — you guessed it — 3- or 4-inch pieces and scatter them around. Don't just drop them where you cut them; this looks messy. Scatter them to make them part of a mixed mulch, which looks tidy.

When you have bigger, coarser branches that you can't easily cut up, create a designated spot to put them. Be sure that the branches lie flat on the ground by making them two-dimensional, ideally by cutting off all side branches, but at a minimum by cutting off any twigs that stick up or down. If the branches aren't touching the ground, they won't rot, and you'll end up with a dry, brushy mess. Throw hay and garden waste over the flattened material to hasten the rotting process and make a good, diverse compost. When you're ready to plant, pull out whatever hasn't rotted, move it to the next spot, and plant in the soil that's been enriched by all that mulch.

When You're Serious: Machete Maintenance

At the less slothful end of the maintenance spectrum is working with a machete. A few times a year, when my garden starts to look scruffy because vegetable plants have run their course and shrubs need pruning, I do a major cleanup. This is serious work for a day when you want a workout and psychological catharsis. With my approach, a major cleanup is kind of fun — exhilarating, even. For this adventure, wear sturdy shoes, long pants, and long sleeves. Then grab your pruning shears, holster, and machete.

In addition to releasing aggressions and making you look badass, a machete makes short work of unsightly plant matter. Pull up spent vegetables and flowers, hold them by the roots, and hack them to bits with the machete, facing in different directions to keep the mulch mixed and thus presentable (the exception is roses; see page 154).

ROSES: THE EXCEPTION, AGAIN

One exception to the "hold and hack" approach to pruning is rosebushes. If you try to hold them while you hack with the machete, they'll stab the dickens out of you, possibly causing you to take the Lord's name in vain. Instead, carefully find a place between thorns and hold the branch upside down, then cut it into short pieces with clippers, working up from the tip. Scatter the pieces more closely at the base of the rosebush, so you know where the thorns are when you weed. Later, remember not to do two-fisted weeding under rosebushes; use a hand hoe instead, or gingerly weed with two fingers.

It also often suffices to just break off dead rose blossoms and scatter the petals around. This nonpurist approach alone will keep roses looking pretty darned good.

This process is especially satisfying with dead cornstalks and squash plants; their removal causes a huge leap in garden presentability while adding lots of organic matter to the mulch.

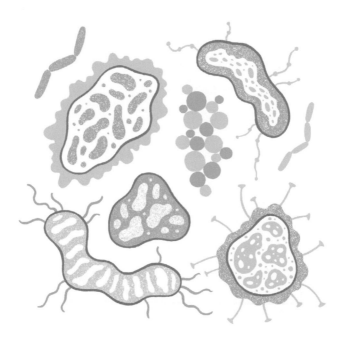

So, now you're acquainted with the universe that lives in your soil. You know how to plant a garden and how to keep it healthy and tidy. Now what?

Go out there and plant! And plant, and plant! Plus, mulch. As you add more plants and food on top of the soil, in the form of compost and mulch, your soil will become healthier and healthier. Also, think about who's growing the food you buy. Are you buying from a CSA (community supported agriculture)? Are they using organic methods? Can you convince them not to till? Let's start a trend and try to save the planet, piece by piece. Let's do better.

I love to hear from readers — please feel free to contact me with questions or feedback on my website: dianemiessler.com.

GLOSSARY

ALLELOPATHIC (a-LEL-o-PATH-ic). Growth-suppressing. Pine needles and walnut leaves are allelopathic, which is why not much grows under those trees.

ANION (an-EYE-un). An ion with a negative electrical charge, because it has an extra electron or two.

CARBON SEQUESTRATION. The process of pulling carbon out of the air and changing it into solid or liquid form, as opposed to sending it into the atmosphere to make carbon dioxide and destroy life as we know it. This can be a fast or slow process. Sequestration is done by all lifeforms and their remains, including fossil fuels. The ocean also absorbs carbon dioxide, some of which remains in liquid form or is used to make ocean creatures and their shells.

CARBON SINK. A reservoir that holds carbon; the major carbon reservoirs on Earth are oceans, fossil fuels, soil, and the atmosphere. That last one is where we run into problems.

CATION (CAT-eye-un). An ion with a positive electrical charge; i.e., a below-average number of electrons.

CATION EXCHANGE CAPACITY (CEC). A measure of how many cations (positively charged ions, like potassium, ammonia, and calcium) can be retained on soil particle surfaces. CEC is used as a measure of soil fertility, since it indicates the ability of the soil to retain certain nutrients.

ECTOMYCORRHIZAE (eck-toe-my-co-RI-zee). Long fungal strands that twine around roots.

ENDOMYCORRHIZAE (en-doe-my-co-RI-zee). Long fungal strands that grow into roots.

EUKARYOTE (you-CARE-e-oat). An organism made up of cells whose DNA is enclosed within a nucleus, not just floating around willy-nilly in the cell like in the more primitive prokaryotes.

Eukaryotes can reproduce either by division (where the offspring are identical to the parent) or sexually, where the DNA from two different cells combines to make something entirely different.

Eukaryotes include everything but bacteria and archaea — everything from fungi on up to humans. They happened when life on Earth started to get organized. About half a billion years ago, eukaryotes started banding together to form the precursors of plants and animals.

HYPHAE (HI-fee; SINGULAR: HYPHA). Long strings of single fungal cells, invisible to the naked eye.

ION. An atom or molecule (cluster of atoms) that has an electrical charge due to the loss of one or more electrons (little things that move around atoms). Plant nutrients like N-P-K (nitrogen, phosphorus, and potassium) are ions.

MICROBIOME. An ecological community of microorganisms found in and on all multicellular organisms studied to date, from plants to animals. Microbiomes have been found to be crucial for the immunologic, hormonal, and metabolic function of their host. All plants and animals, from simple life-forms to humans, live in close association with microbial communities.

MICROORGANISM/MICROBE. Something alive you can't see without a microscope; this includes bacteria, fungi, archaea, some algae, and lots of other very, *very* small things, some of which we don't know about yet, because science is always learning. Science is like that.

MULCH. A layer of organic matter that preserves moisture, smothers weeds, feeds soil life, then breaks down into compost, all while looking somewhat-to-highly attractive.

MYCELIUM. A bundle of hyphal strands; you can see these.

MYCORRHIZAE (my-co-RI-zee; SINGULAR: MYCORRHIZA). Long fungal strands (mycelium) that grow around or into plant roots and feed them.

ORGANIC. Of, related to, or derived from living things. Carbon-containing.

ORGANIC MATTER. Anything that is currently or formerly alive, including microbes. For our composting purposes, this will mostly include vegetable scraps from your kitchen, dead leaves, grass clippings, and spent plants from the garden.

PROKARYOTE (pro-CARE-e-oat). Usually a single-celled organism, one whose DNA is just floating around in the cellular fluid. Prokaryotes are probably the first form of life on Earth and reproduce only by division. They include archaea and bacteria; everything else on Earth is made of eukaryotes.

The prokaryote cyanobacteria, sometimes called blue-green algae, was the first organism to figure out how to photosynthesize, and it paved the way for the existence of plants. And us.

RHIZOSPHERE (RI-zo-sphere). The area around plant roots; the root zone. If you play your cards right, the rhizosphere is some very rich soil.

SOIL. As defined by me, and science, the outermost layer of the planet where the underlying rock and clay interacts with the outside world. This happy marriage of inert minerals with air, water, and living organisms (germs, plants, and animals, to name a few) creates a carbon-rich, crumbly, biologically active substance that grows large and delicious tomatoes. Just to clarify something: that bag of stuff you buy labeled "potting soil" is not soil.

SOIL FOOD WEB. A term coined by soil scientist Dr. Elaine R. Ingham. It refers to all the things that live in, and right on top of, soil, most of which eat or are eaten by each other and, in the process, create rich soil.

BIBLIOGRAPHY

Appelhof, Mary, and Joanne Olszewski. *Worms Eat My Garbage*. Storey Publishing, 2017.

Bartocci, Pietro, et al. "Effect of Biochar on Water Retention in Soil, A Comparison Between Two Forms: Powder and Pellet." ResearchGate, June 2017. https://www.researchgate.net/publication/317642821 _EFFECT_OF_BIOCHAR_ON_WATER_RETENTION_IN_SOIL_A _COMPARISON_BETWEEN_TWO_FORMS_POWDER_AND_PELLET.

Biello, David. "The Origin of Oxygen in Earth's Atmosphere." *Scientific American*, August 19, 2009. https://www.scientificamerican.com /article/origin-of-oxygen-in-atmosphere.

Bonanomi, G., F. Ippolito, and F. Scala. "A 'black' future for plant pathology? Biochar as a new soil amendment for controlling plant diseases." *Journal of Plant Pathology* 97, no. 2 (2015): 223–34.

Briscoe, Charles B. "Early Results of Mycorrhyzal Inoculation of Pine in Puerto Rico." Tropical Forest Research Center, USDA, July–December 1959. https://www.srs.fs.usda.gov/pubs/ja/1959/ja_1959_briscoe_001.pdf.

"CEC and Holding N in the Soil." Agvise Laboratories. https://www.agvise .com/educational-articles/cec-and-holding-n-in-the-soil.

Gibson, Rick. "Cations and Anions in the Soil." Pinal Central, July 25, 2017. https://www.pinalcentral.com/home_and_hearth/gibson-cations-and -anions-in-the-soil/article_1697050f-86a6-52d7-95eb-5725c1d2784b.html.

Gilker, Rachel. "Soil chemistry? Are you kidding? CEC in Under 500 Words." On Pasture, August 12, 2013. https://onpasture.com/2013/08/12/soil -chemistry-are-you-kidding-cec-in-400-words.

Hacskaylo, Edward. "Mycorrhizae: Proceedings of the First North American Conference on Mycorrhizae." University of Illinois at Urbana Champaign, April 1969.

Hess, Anna. *Homegrown Humus: Cover Crops in a No-Till Garden*. Wetknee Books, 2013.

Ingham, Elaine R., Andrew R. Moldenke, and Clive A. Edwards. *The Soil Biology Primer*. National Resources Conservation Service, 2000.

Kopecky, Mark. "When to Use Lime, Gypsum, and Elemental Sulfur." On Pasture, June 2, 2014. https://onpasture.com/2014/06/02/when-to-use-lime-gypsum-and-elemental-sulfur.

Lowenfels, Jeff and Wayne Lewis. *Teaming with Microbes*. Timber Press, 2010.

McGrath, Mike, Howard Garrett, and Lee Reich. "Can You — SHOULD You — Compost Diseased Tree Leaves?" Gardens Alive, February 5, 2001. https://www.gardensalive.com/product/ybyg-can-you-should-you-compost-deseased-tree-leaves.

Melendrez, Michael M. "Humic Acid: The Science of Humus and How It Benefits Soil." Ecofarming Daily, August 2009. http://www.ecofarmingdaily.com/humic-acid.

Mengel, David B. "Fundamentals of Soil Cation Exchange Capacity (CEC)." *Agronomy Guide*, Purdue University Cooperative Extension Service, March 1993. https://www.extension.purdue.edu/extmedia/ay/ay-238.html

Ohlson, Kristin. 2014. *The Soil Will Save Us*. Rodale Books, 2014.

Rosatto, Carl, Helen Atthowe, and Alex Stone. "Woodleaf Farm Soil Management System." Oregon State University Extension, May 31, 2019. https://articles.extension.org/pages/73427/woodleaf-farm-soil-management-system.

Stamets, Paul. *Mycelium Running*. Ten Speed Press, 2005.

Vozzo, J. A., and Edward Hacskaylo. "Inoculation of Pinus Caribaea with Ectomycorrhizal Fungi in Puerto Rico." *Forest Science* 17, no. 2 (June 1971): 239–45.

Waksman, Selman. *Humus: Origin, Chemical Composition, and Importance in Nature*. The Williams and Wilkins Company, 1936.

METRIC CONVERSIONS

LENGTH

To convert	to	multiply
inches	millimeters	inches by 25.4
inches	centimeters	inches by 2.54
inches	meters	inches by 0.0254
feet	meters	feet by 0.3048
feet	kilometers	feet by 0.0003048
yards	centimeters	yards by 91.44
yards	meters	yards by 0.9144
yards	kilometers	yards by 0.0009144
miles	meters	miles by 1,609.344
miles	kilometers	miles by 1.609344

AREA

To convert	to	multiply
acres	hectares	acres by 2.47

WEIGHT

To convert	to	multiply
pounds	grams	pounds by 453.5
pounds	kilograms	pounds by 0.45

VOLUME

To convert	to	multiply
teaspoons	milliliters	teaspoons by 4.93
tablespoons	milliliters	tablespoons by 14.79
quarts	milliliters	quarts by 946.36
quarts	liters	quarts by 0.946
gallons	liters	gallons by 3.785

INDEX

Page numbers in *italic* indicate images.

Actinomycetes, 46, 47, 56
aeration, 74–75
aggregates, 44, 55
alfalfa hay, 26
algae, 62
alyssum, 39
amoebae, 63
anecic worms, 67
anions, 77
antibiotics, 46, 47, 56
arachnids, 65
archaea, 51, 60–61
arthropods, 65
asbestos, 12
asexual reproduction, 54

Bacteria, 23, 54–56, 119–120. *See also* microbes
bacterial slime, 55
beans, 35
beer traps, 68
Bermuda grass, 116
biochar, 78–83, *81*
biodiversity
   algae, bugs, worms, etc. and, 62–69
   microbes and, 46–61
biofilms, *48*
bioremediation, 61
black soldier flies (*Hermetia illucens*), 126, 127
boron, 103
broadleaf plant cover crops, 27
buckwheat, 28–29, 33, 34
Bushby, H. V. A., 6
bush peas, 29

Calcium, 88, 92–93, *92*, 100
calories, 86–87
carbon, organic matter and, 42
carbon dioxide, 16–19, *48*, *86*, 87
carbon sequestration, 16–19, 73
cardboard mulch, 24–25
castings, 67, 121
cation exchange capacity (CEC), 7, 71, 77–83, 98, 104
cations, 77, 103
cellulose, 21, 86–87
centipedes, 65
charcoal, 79–83
chemical fertilizers, 4, 14–15, 78, 89
chlorophyll, *48*, 87, 89, 93
clay, 11–12, 78, 90
clippers, 142
Clostridium difficile, 53
clover, 34
Cobb, Nathan, 63
coffee grounds, 68
composition of soil, 10–11
compost
   benefits of, 108–109
   building pile, 111–113
   fungal vs. bacterial dominance of, 119–120
   maintaining pile, 113, 114–117
   mulch vs., 110
   overview of, 107
   troubleshooting, 118–119, 123–127
   when to use, 117
compost bins, 115
compost tea, 128–131
compost tumblers, 115
conductivity, 101
cover crops, 4, 27–39. *See also specific plants*
cowpeas, 35

crumbles, 73
crumbs, 44
cyanobacteria, 54

Daikon, 30, 33, 34
deadheading, 150-151
decomposition, 50, 55
diatomaceous earth, 62, 68, 69
diffusion, 94
digestion, 50-51
diseases, 56, 149. See also pathogens
Dust Bowl, 74

Earthworms, 66, 67
ECe. See electrical conductivity
ectomycorrhizae, 60
edamame, 29
edible cover crops, 35, 39
Eisenia fetida, 67
electrical conductivity (ECe), 101
endogeic worms, 65
endomycorrhizae, 60
epigeic worms, 65, 67
eukaryotes, 56, 62
extremophiles, 60-61

Fall cover crops, 34
fava beans, 30, 34
fecal transplants, 52
fertilizers. See chemical fertilizers
Fiskars clippers, 142
Fiskars machete, 143
flies, 126, 127
food web (soil), 42, 43, 44-46, 45
forage radishes, 30, 33, 34
fungi, 23, 56-60, 90, 119-120

Global warming, 16-19
gloves, 143
grass cover crops
    overview of, 27
    planting of, 33, 34, 35
    specific, 28-29, 30
grasses, 116

gravel, 76
green beans, 29
green manure, 149
gypsum, 12, 92-93, 92, 139

Harlequin bugs, 69
hay mulch, 26
hemimastigotes, 51
Hemimastix kukwesjijk, 51
hermaphrodites, 66
Hess, Anna, 31, 77
Huckenpahler, B. J., 59
humus, 21, 23, 42
hyphae, 56, 57

Jones, Christine, 6

Ken Ho hoes, 142
kilns, 82

Legume cover crops, 27, 29-30, 35
Lewis, Wayne, 57
Libby Mine (Montana), 12
lignins, 21
lime, 104
Lowenfels, Jeff, 57

Machetes, 143, 153-154
macronutrients, 87, 88-93, 98-102, 105
maggots, 126, 127
magnesium, 88, 93, 100
maintenance, 146-155
microbes
    bacteria, 23, 54-56, 119-120
    compost and, 119-120, 124
    fungi, 23, 56-60, 90, 119-120
    importance of, 50-52
    mulch and, 23
    overview of, 46-48
    reproduction of, 54
    rototilling and, 72
    soil composition and, 13
    three types of, 52
micronutrients, 87, 94, 103, 105

mildew, 56
millipedes, 65
mineralization, 48, 64
minerals, 10, 13. *See also specific minerals*
mollusks, 67–68
mulch
  building garden and, 137
  cardboard, 24–25
  compost vs., 110
  green and brown, 23
  as maintenance, 146
  miscellaneous types of, 26
  overview of, 21, 22
  straw and hay, 26
  turning cover crops into, 36–38
  unsuitable materials for, 22
mulching, 49
mushrooms, 56, 57, 61
mustard, 35
mycelia, 56, 57
mycorrhizae, 4, 17, 57, 58–60, 72

Nasturtiums, 35, 39
nectars, 86–87
nematodes, 63–64
Nematodes and Their Relationships (Cobb), 63
nightcrawlers, 67
nitrogen, 55, 88–89, 98–99
nitrogen fixation, 27, 29–30, 46, 62, 89
N-P-K, 88
nutrients. *See also specific nutrients*
  bacteria and, 55–56
  cation exchange capacity and, 71, 77
  fixing soil and, 105
  food web and, 44
  macronutrients, 87, 88–93, 98–102, 105
  microbes and, 46, 47, 49, 61
  micronutrients, 87, 94, 103, 105
  nematodes and, 64

  overview of, 85
  photosynthesis and, 86–87
  remediation and, 61
  testing for, 95–104

Oats, 30
oilseed radishes, 30, 33, 34
organic matter. *See* soil organic matter
osmosis, 94
oxidization, 18
oyster mushrooms, 61

Pathogens, 46, 52, 53, 128–129
paths, 75–76, 136
peas, 34, 35, 39
percent cation saturation, 103, 105
perfectionism, 151
perlite, 12
pH
  calcium and, 92–93, *92*
  compost and, 120
  fixing soil and, 105
  phosphorus and, 90
  pine needles and, 138
  testing for, 96–97
phosphorus, 88, 90–91, 99
photosynthesis, 46, 48, 86–87, *86*
pill bugs, 65
pine needles, 138
pine trees, 58–59
planting, 138–141
poppies, 35, 39
potassium, 88, 91, 99
prokaryotes, 56, 62
protozoa, 63
pruning, 152–153
pruning shears, 146–147
Puerto Rican pines, 58–59
puttering, 146–153

Quality of soil, 11–15

Radishes, oilseed, 30, 33, 34
red wigglers, 67

Reich, Lee, 152–153
remediation, 61
retort kilns, 82
rhizosphere, 48
root hairs, 49
roots, 5, 48–49, 48, 94, 139–140
root zone, 15
roses, 151, 154
rototilling
   carbon and, 18
   commandment against, 4
   food web and, 46
   mycorrhizae and, 58
   reasons to avoid, 72–74
   soil organic matter and, 42, 43
rotting, 50
roundworms, 63–64
ryegrass, 30

Sand, 11–12
seeds, 26, 149
sexual reproduction, 54
shredders, 143
shrubs, 139–140
silt, 11–12
slugs, 67–68
smothering, 38
snails, 67–68
socks, 143
sodium, 100
soil organic matter (SOM)
   algae, bugs, worms, etc. and, 62–69
   cation exchange capacity and, 77–78
   food web and, 15, 44–46, 45
   microbes and, 46–61
   overview of, 14, 41–43
   percentage of, 10
   testing for, 98
sow bugs, 65
soybeans, 29–30, 35
spinach, 39
spores, 57
spring cover crops, 35

Stamets, Paul, 61
Stout, Ruth, 26
strawberries, 35, 39
straw mulch, 26
Sudan grass, 35
sugars, 48, 86–87
suggestions, overview of, 4–5
sulfur, 88, 93, 105
summer cover crops, 35
sunflowers, 35
sweat, 46
sweet potatoes, 31–32, 35, 39

Teaming with Microbes (Lowenfels and Lewis), 57
terra preta, 79
testing of soil, 95–104
tilling
   carbon and, 18
   commandment against, 4
   food web and, 46
   mycorrhizae and, 58
   reasons to avoid, 72–74
   soil organic matter and, 42, 43
tools, 142–143
trace elements (micronutrients), 87, 94, 103, 105
transpiration, 93
trees, 139–140, 152
Turnbull, Cass, 152–153

Vermiculite, 12
vermiculture, 123
vetch, 34
viruses, 53

Watering systems, 134–135, 143
weak bray, 99
weathering, 11
weeding, 147–149
weeds, 5, 124
wood chips, 76
worm bins, 121–122, *121*
worms, 65–67, *66*, 72
worm tea, 123

Cultivate Gardening Success
with More Books from Storey

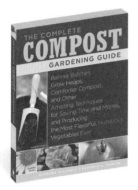

The Complete Compost Gardening Guide
by Barbara Pleasant & Deborah L. Martin

This thorough, informative tour of materials and innovative techniques helps you turn an average vegetable plot into a rich incubator of healthy produce and an average flower bed into a rich tapestry of bountiful blooms all season long.

Let It Rot!
by Stu Campbell

Stop bagging leaves, grass, and kitchen scraps, and turn household waste into the gardener's gold: compost. This classic guide offers accessible advice for starting and maintaining a composting system, building bins, and using your finished compost.

Worms Eat My Garbage, 35th Anniversary Edition
by Mary Appelhof & Joanne Olszewski

Use worms to recycle food waste into nutrient-rich fertilizer for your houseplants or garden. This beloved guide teaches you everything you need to know to buy or build your own worm bin, maintain your worms, and harvest the compost.

Join the conversation. Share your experience with this book, learn more about Storey Publishing's authors, and read original essays and book excerpts at storey.com. Look for our books wherever quality books are sold or call 800-441-5700.